TWO THOUSAND MILES BELOW

By

CHARLES WILLARD DIFFIN

Two Thousand Miles Below
by Charles Willard Diffin

ISBN: 978-93-60468-18-7

Published by

DOUBLE 9 BOOKS

2/13-B, Ansari Road, Daryaganj
New Delhi – 110002
info@double9books.com
www.double9books.com
Tel. 011-40042856

ABOUT THE AUTHOR

Charles Willard Diffin (1884-1966) was an American novelist, engineer, and salesperson who created a name for himself in the field of science fiction. While Diffin experimented in other genres such as Westerns and mysteries, he is most known for his contributions to science fiction. His writing career began in 1930, when he made his debut in the groundbreaking science fiction journal Astounding Stories. Diffin's science fiction works stand out for their innovative explorations of futuristic notions, space travel, and technical advances. His paintings frequently reflected the spirit of the early twentieth century's curiosity with science and the unknown. He took readers to faraway planets, introduced them to strange cultures, and challenged traditional thinking with speculative notions in his writings. Diffin offered a rare blend of technical aptitude and storytelling flair to his job as both an engineer and a salesman. This enabled him to create storylines that were not only captivating but also scientifically plausible, bringing depth and credibility to his science fiction stories. The enduring legacy of Charles Willard Diffin is his ability to combine the realms of science and imagination, enthralling readers with his visions of the future while making an unmistakable stamp on the early science fiction landscape.

CONTENTS

CHAPTER I
A MAN NAMED SMITH

Heat! Heat of a white-hot sun only two hours old. Heat of blazing sands where shimmering, gassy waves made the sparse sagebrush seem about to burst into flames. Heat of a wind that might have come out of the fire-box of a Mogul on an upgrade pull.

A highway twisted among black masses of outcropping lava rock or tightened into a straightaway for miles across the desert that swept up to the mountain's base. The asphalt surface of the pavement was almost liquid; it clung stickily to the tires of a big car, letting go with a continuous, ripping sound.

Behind the wheel of the weatherbeaten, sunburned car, Dean Rawson squinted his eyes against the glare. His lean, tanned face was almost as brown as his hair. The sun had done its work there; it had set crinkly lines about the man's eyes of darker brown. But the deeper lines in that young face had been etched by responsibility; they made the man seem older than his twenty-three years, until the steady eyes, flashing into quick amusement, gave them the lie.

And now Rawson's lips twisted into a little grin at his own discomfort— but he knew the desert driver's trick.

"A hundred plus in the shade," he reasoned silently. "That's hot any way you take it. But taking it in the face at forty-five an hour is too much like looking into a Bessemer converter!"

He closed the windows of his old coupe to within an inch of the top, then opened the windshield a scant half inch. The blast that had been drawing the moisture from his body became a gently circulating current of hot air.

He had gone only another ten miles after these preparations for fast driving, when he eased the big weatherbeaten car to a stop.

On his right, reaching up to the cool heights under a cloudless blue sky, the gray peaks of the Sierras gave promise of relief from the furnace breath of the desert floor. There were even valleys of snow glistening whitely where the mountains held them high. A watcher, had there been one to observe in the empty land, might have understood another traveler's pausing to admire the serene majesty of those heights — but he would have wondered could he have seen Rawson's eyes turned in longing away from the mountains while he stared across the forbidding sands.

There were other mountains, lavender and gray, in the distance. And nearer by, a matter of twenty or thirty elusive miles through the dancing waves of hot air, were other barren slopes. Across the rolling sand-hills wheel marks, faint and wind-blown, led straight from the highway toward the parched peaks.

"Tonah Basin!" Rawson was thinking. "It's there inside these hills. It's hotter than this is by twenty degrees right this minute — but I wish I could see it. I'd like to have one more look before I face that hard-boiled bunch in the city!"

He looked at his watch and shook his head. "Not a chance," he admitted. "I'm due up in Erickson's office in five hours. I wonder if I've got a chance with them...."

Five hours of driving, and Rawson walked into the office of Erickson, Incorporated, with a steady step. Another hour, and his tanned face had gone a trifle pale; his lips were set grimly in a straight line that would not relax under the verdict he felt certain he was about to hear.

For an hour he had faced the steely-eyed man across the long table in the Directors Room — faced him and replied to questions from this man and the half-dozen others seated there. Skeptical questions, tricky questions; and now the man was speaking:

"Rawson, six months ago you laid your Tonah Basin plans before us — plans to get power from the center of the Earth, to utilize that energy, and to control the power situation in this whole Southwest. It looked like a wild gamble then, but we investigated. It still looks like a gamble."

"Yes," said Rawson, "it is a gamble. Did I ever call it anything else?"

"The Ehrmann oscillator," the man continued imperturbably, "invented in 1940, two years ago, solves the wireless transmission problem, but the success of your plan depends upon your own invention — upon your straight-line drills that you say will not wander off at a tangent when they get down a few miles. And more than that, it depends upon you.

"Even that does not damn the scheme; but, Rawson, there's only one factor we gamble on. No wild plans, no matter how many hundreds of millions they promise: no machines, no matter what they are designed to do, get a dollar of our backing. It's men we back with our money!"

Rawson's face was set to show no emotion, but within his mind were insistent, clamoring thoughts:

"Why can't he say it and get it over with? I've lost — what a hard-boiled bunch they are! — but he doesn't need to drag out the agony." But — but what was the man saying?

"Men, Rawson!" the emotionless voice continued. "And we've checked up on you from the time you took your nourishment out of a bottle; it's you we're backing. That's why we have organized the little company of Thermal Explorations, Limited. That's why we've put a million of hard coin into it. That's why we've put you in charge of operations."

He was extending a hand that Dean Rawson had to reach for blindly.

"I'd drill through to hell," Dean said and fought to keep his voice steady, "with backing like that!"

He allowed his emotion to express itself in a shaky laugh. "Perhaps I will at that," he added: "I'll certainly be heading in the right direction."

Under another day's sun the hot asphalt was again taking the print of the tires of Rawson's old car. But this time, when he came to the almost obliterated marks that led through the sand toward distant mountains, he stopped, partially deflated the tires to give them a grip on the sand, and swung off.

"A fool, kid trick," he admitted to himself, "but I want to see the place. I'll see plenty of it before I'm through, but right now I've got to have a look; then I'll buckle down to work.

"Thermal Explorations, Limited!" The name rang triumphantly in his mind. "A million things to do — men, crews for the drills, derricks.... We'll have to truck in over this road; I'll lay a plank road over the sand. And water — we'll have to haul that, too, until we can sink a well. We'll find water under there somewhere. I've got to see the place...."

The black sides of the mountains were nearer: every outcropping rock was plainly volcanic, and great sweeping slopes were beds of ash and pumice; the wheel marks, where they showed at all, wound off and into a canyon hidden in the tremendous hills that thrust themselves abruptly from the desert floor.

The mountains themselves towered hugely at closer range, but the road that Rawson followed climbed through them without traversing the highest slopes. It was scarcely more than a trail, barely wide enough for the car at times, but boulder-filled gullies showed where the hands of men had worked to build it.

He came at last into the open where a shoulder of rock bent the road outward above a sea of sand far below. And now the mountains showed their circular arrangement—a great ring, twenty miles across. At one side were three conical peaks, unmistakable craters, whose scarred sides were smothered under ash and sand that had rained down from their shattered tops in ages past. Yet, so hot they were, so clear-cut the irregularly rimmed cups at their tops, that they seemed to have pushed themselves up through the earth in that very instant. At their bases were signs of human habitation—broken walls, scattered stone buildings whose empty windows gaped blackly. This was all that remained of New Rhyolite.

Rawson looked at the "ghost town" which had never failed to interest him, but he gave no thought now to the hardy prospectors who had built it or to the vein of gold that had failed them. His searching eyes came back to the fiery pit, the Tonah Basin, a vast cauldron of sand and ash—great sweeps of yellow and gray and darker brown into which the sun was pouring its rays with burning-glass fierceness.

But to Rawson, there was more than the eye could see. He was picturing a great powerhouse, steel derricks, capped pipes that led off to whirring turbines, generators, strings of cables stretching out on steel supports into the distance, a wireless transmitter—and all of this the result of his own vision, of the stream he would bring from deep in the earth!

Then, abruptly, the pictures faded. Far below him on the yellow, sun-blasted floor, a fleck of shadow had moved. It appeared suddenly from the sand, moved erratically, staggeringly, for a hundred feet, then vanished as if something had blotted it out—and Dean Rawson knew that it was the shadow of a man.

The road widened beyond the turn. He had intended to swing around; he had wanted only to take a clear picture of the place with him. But now the big car's gears wailed as he took the downgrade in second, and the brakes, jammed on at the sharp curves, added their voice to the chorus of haste.

"Confounded desert rats!" Rawson was saying under his breath. "They'll chance anything—but imagine crossing country like that! And he hasn't a burro—he's got only the water he can carry in a canteen!"

But even the canteen was empty, he found, when he stopped the car in a whirl of loose sand beside a prone figure whose khaki clothes were almost indistinguishable against the desert soil.

Before Rawson could get his own lanky six feet of wiry length from the car, the man had struggled to his feet. Again the little blot of shadow began its wavering, uncertain, forward movement.

He was a little shorter than Rawson, a little heavier of build, and younger by a year or two, although his flushed face and a two days' stubble of black beard might have been misleading. Rawson caught the staggering man and half carried him to the shadow of the car, the only shelter in that whole vast cauldron of the sun.

From a mouth where a swollen tongue protruded thickly came an agonized sound that was a cry for, "Water—water!" Rawson gave it to him as rapidly as he dared, until he allowed the man to drink from the desert bag at the last. And his keen eyes were taking in all the significant details as he worked.

The khaki clothes earned a nod of silent approval. The compact roll that had been slung from the younger man's shoulders, even the broad shoulders themselves, and the square jaw, unshaved and grimy, got Rawson's inaudible, "O. K.!" But the face was more burned than tanned.

He introduced himself when the stranger was able to stand. "I'm Rawson, Dean Rawson, mining engineer when I'm working at it," he explained. "I'm bound north. I'll take you out of this. You can travel with me as far as you please."

The dark-haired youngster was plainly youthful now, as he stood erect. His voice was recovering what must have been its usual hearty ring.

"I'm not trying to say 'thank you,'" he said, as he took Rawson's hand. "I was sure sunk—going down for the last time—taps—all that sort of thing! You pulled me out—the good old helping hand. Can't thank a fellow for that—just return the favor or pass it on to someone else. And, by the way— you won't believe it—but my name is Smith."

Rawson smiled good-naturedly. "No," he agreed, "I don't believe it. But it's a good, handy name. All right, Smithy, jump in! Here, let me give you a lift; you're still woozy."

Rawson found his passenger uncommunicative. Not but what Smithy talked freely of everything but himself, but it was of himself that Rawson wanted to know.

"Drop me at the first town," said Smithy. "You're going north: I'm south-bound—looking for a job down in Los. I won't take any more short cuts; I was two days on this last one. I'll stick to the road."

They were through the mountains that ringed in the fiery pit of Tonah Basin. Smooth sand lay ahead; only the shallow marks that his own tires had ploughed needed to be followed. Dean Rawson turned and looked with fair appraisal at the man he had saved.

"Drifter?" he asked himself silently. "Road bum? He doesn't look the part; there's something about him...."

Aloud he inquired: "What's your line? What do you know?"

And the young man answered frankly: "Not a thing!"

Dean sensed failure, inefficiency. He resented it in this youngster who had fought so gamely with death. His voice was harsh with a curious sense of his own disappointment as he asked:

"Found the going too hard for you up north, did you? Well, it won't be any easier—" But Smithy had interrupted with a weak movement of his hand.

"Not too hard," he said laconically; "too damn soft! I don't know what I'm looking for—pretty dumb: got a lot to learn!—but it'll be a job that needs to take a good licking!"

"'Too damn soft!'" Dean was thinking. "And he tackled the desert alone!" There was a lot here he did not understand. But the look in the eyes of Smithy that met his own searching gaze and returned it squarely if a bit whimsically—that was something he *could* understand. Dean Rawson was a judge of men. The sudden impulse that moved him was founded upon certainty.

"You've found that job," he said. "The desert almost got you a little while ago—now it's due to take that licking you were talking about. I'm going to teach it to lie down and roll over and jump through hoops. Fact is, my job is to get it into harness and put it to work. I'll be working right out there in the Basin where I found you. It will be only about two degrees cooler than hell. If that sounds good to you, Smithy, stick around."

He warmed oddly to the look in the younger man's deep-set, dark eyes, as Smithy replied:

"Try to put me out, Rawson—just try to put me out!"

CHAPTER II
GOLD!

"Ten miles down, drillers!
Hell-bound, and proud of it!
Ten miles down, drillers!
Hark to what I say:
You're pokin' through the crust of hell
And braggin' too damn loud of it,
For, when you get to hell, you'll find
The devil there to pay."

From the black, night-wrapped valley, far below, the singer's voice went silent with the slamming of a door in one of the bunkhouses. The song was popular; some rimester in the Tonah Basin camp had written the parody for the tormenting of the drill crews. And, high on the mountainside, Dean Rawson hummed a few bars of the lilting air after the singer's voice had ceased.

"Ten miles down!" he said at last to his assistant, sprawled out on the stone beside him. "That's about right, Smithy. And maybe the rest of the doggerel isn't so far off either. 'Pokin' through the crust of hell' — well, there was hell popping around here once, and I am gambling that the furnaces aren't all out."

They were on the outthrust shoulder of rock where the mountain road hung high above the valley floor. Below, where, months before, Rawson had rescued a man from desert death, was blackness punctured by points of light — bunkhouse windows, the drilling-floor lights at the foot of a big derrick, a single warning light at the derrick's top. But the buildings and the towering steelwork of the derrick that handled the rotary drills were dim and ghostly in the light of the stars.

"We've gone through some places I'd call plenty warm," said Smithy, "but you — you craves it *hot*! Think we're about due?" he asked.

Rawson answered indirectly.

"One great big old he-crater!" he said. His outstretched arm swept the whole circle of starlit mountains that enclosed the Basin. "That's what this was once. Twenty miles across — and when it blew its head off it must have sprayed this whole Southwest.

"Now, those craters" — he pointed contemptuously toward the three conical peaks off to the right — "those were just blow-holes on the side of this big one."

In the ragged ring of mountains, the throat of some volcanic monster of an earlier age, the three cones towered hugely. Their tops were plainly cupped; their ashy sloping sides swept down to the desert floor. At their base, the gray walls of stone in the ghost town of Little Rhyolite gleamed palely, like skeleton remains.

"I've seen steam, live steam," Rawson went on, "coming out of a fissure in the rocks. I know there's heat and plenty of it down below. We're about due to hit it. The boys are pulling the drill now; they cut through into a whale of a cave down below there — "

He broke off abruptly to fix his attention on the dark valley below, where lights were moving. One white slash of brilliance cut across the dark ground; another, then a cluster of flood lights blazed out. They picked the skeleton framework of the giant derrick in black relief against the white glare of the sand. From far below; through the quiet air, came sounds of excited shouting; the voices of men were raised in sudden clamor.

"They've pulled the drill," said Rawson. "But why all the excitement?"

He had already turned toward their car when the crackle of six quick shots came from below. His abrupt command was not needed; Smithy was in the car while still the echoes were rolling off among the hills. Their own lights flashed on to show the mountain grade waiting for their quick descent.

The sandy floor of this part of the Tonah Basin was littered with the orderly disorder of a big construction job — mountains of casing, tubular drill rod, a foot in diameter; segmental bearings to clamp around the rod every hundred feet and give it smooth play. Dean drove his car swiftly along the surfaced road that was known as "Main Street" to the entire camp.

There were men running toward the derrick — men of the day shift who had been aroused from their sleep. Others were clustered about the wide concrete floor where the derrick stood. Clad only in trousers and shoes, their bodies, tanned by the desert sun, were almost black in the glare of the big floods. They milled wildly about the derrick; and, through all their clamor and shouting, one word was repeated again and again:

"Gold! Gold! Gold!"

The big drill head was suspended above the floor. Dean Rawson, with Smithy close at hand, pushed through the crowd. He was prepared to see traces of gold in the sludge that was bailed out through the hollow shaft—quartz, perhaps, whose richness had set the men wild before they realized how impossible it would be to develop such a mine. But Rawson stopped almost aghast at the glaring splendor of the golden drill hanging naked in the blinding light.

Riley, foreman of the night shift, was standing beside it, a pistol in his hand. "L'ave it be," he was commanding. "Not a hand do ye lay on it till the boss gets here." At sight of Rawson he stepped forward.

"I shot in the air," he explained. "I knew ye were up in the hills for a breath of coolness. I wanted to get ye here quick."

"Right," said Rawson tersely. "But, man, what have you done with the drill? It's smeared over with gold!"

"Fair clogged wid it, sir," Riley's voice betrayed his own excitement. "You remimber we couldn't pull it at first—the drill was jammed-like after it bruk through at the ten-mile livil. Then it come free—and luk at it! Luk at the damn thing! Sent down for honest work, it was, and it comes back all dressed up in jewelry like a squaw Indian whin there's oil struck on the reservation! Or is it gold ye were after all the time?" he demanded.

"Gold! Gold!" a hundred voices were shouting. Dean hardly heard the voice of the foreman, made suddenly garrulous with excitement. He stared at the big drill head, heaped high with the precious metal. It was jammed into the diamond-studded face of the drill; it filled every crack and crevice, a smooth, solid mass on top of the head and against the stem. A workman had brought a singlejack and chisel; he was prying at a ribbon of the yellow stuff. Riley went for him, gun in hand.

"L'ave it be!" he shouted.

"But, confound it all, Dean," Smithy's voice was saying in a tone of disgust, "I thought we were working on a power plant. Not that a gold mine is so bad; but we can't work it—we can't go down after it at ten miles."

"Gold mine!" Rawson echoed. "I'll say it's a gold mine—but not because of the gold. Do you notice anything peculiar about that, Smithy?"

His assistant replied with a quick exclamation:

"You're right, Dean! I knew there was something haywire with that. Solid chunk—been cast around that stem—melted on. And that means—"

"Heat," said Rawson. "It means we've found what we're after. Give the gold to the men; tell them we'll divide it evenly among them. There's more down there, but there's something better: there's energy, power!"

He snapped out quick orders. "Get the temperature. Drop a recording pyrometer. Let me know at once. There'll be plenty doing now!"

Drill rods and cables, all were made of the newest aluminum alloy. The long tube that held the pyrometer was formed of the same metal. Smithy sent it down to get a recording of the temperatures of that subterranean cave into which their tools had plunged.

He adjusted the recording mechanism himself and stood beside the twenty-inch casing that held back the loose sand from the big bore. Then he watched ten sections of cable, each a mile in length, each heavier than the last, as they went hissing into the earth.

From the cable control shed the voice of Riley was calling the depth.

"Fifty-two thousand." Then by hundreds until he cried: "Fifty-two-seven. We're into the big cave! Now another hundred feet."

The cable was moving slowly. In the middle of Riley's call of "Fifty-two-eight," a jangling bell told that the bottom of the pyrometer carrier had touched.

"Up with it," Smithy ordered. "Make it snappy. We'll see if we've got another cargo of gold."

There was an undeniable thrill in this reaching to a tremendous distance underground, this groping about in a deep-hidden cave, where molten gold was to be found. What had they tapped?—he asked himself. He saw visions of some vast pool of hot, liquid gold. Perhaps Dean would have to change his plans. They could rig up some kind of a bailer; they could bring out thousands of dollars at a time.

He was watching for the first sight of the metal carrier, far more interested in what might be clinging to it than in the record of the pyrometer it held. He saw it emerge—then he stared in disbelief at the stubby mass at the cable's end, where all that remained of the long tube he had sent down was a dangling two feet of discolored metal, warped and distorted. The lower part, a full twenty feet in length, had been fused cleanly off.

Dean Rawson was there to watch the next attempt. Again Riley's roaring bass rolled out the count, but this time the call stopped at fifty-two-seven. The jangling bell told that the carrier had touched.

"Divil a bit do I understand this," Riley was calling. "We're right at the point where we dropped through into the clear. Right at the roof of the big cave—fifty-two-seven, it says—and no lower do we go. The bottom of the hole is plugged!"

Rawson made no reply. He was scowling while he stared speculatively at the mouth of the twenty-inch bore—a vertical tunnel that led from the drilling floor down, down to some inner vault. "Molten gold," he was thinking. "It melted a cylinder of the new Krieger alloy—melted it when its melting point is way higher than that of any rock that we've hit. And now the bore is closed...."

He was trying vainly to project his mental vision through those miles of hard rock to see what manner of mystery this was into which he had probed. He shook his head slowly in baffled speculation, then spoke sharply.

"Drill it out!" he ordered. "We're into a hot spot sure enough, though I can't just figure out the how of it. But we'll tame it, Smithy. Send down the drill. Clean it out. Then we'll poke around down there and get the answer to all this."

Five days were needed to send down the big drill with a new drill-head replacing the other too fouled with gold for any use. The tubular sections, a hundred feet in length, were hooked together and lowered one by one. Each joint meant the coupling of the air-pipe as well. Air, mixed with water from the outer jacket, must come foaming up through the central core to bring the powdered rock to the surface.

Five days, then one hour of boring, and another five days to pull out the drill before Rawson could hope for his answer. But he found it in the severed shaft of the great drill where the head had been melted completely off. The big stem that would have resisted all but electric furnace heat, and been cut through like a tallow candle in the blast of an oxy-acetylene flame.

CHAPTER III
RED DROPS

The flat-roofed shack of yellow boards that was Dean Rawson's "office" had a second canopy roof built above it and extending out on all sides like a wooden umbrella. Thick pitch fried almost audibly from the fir boards when the sun drove straight from overhead, but beneath their shelter the heat was more bearable.

By an open window, where a hot breeze stirred sluggishly, Rawson sat in silent contemplation of the camp. His face was as copper-colored as an Apache's and as motionless. His eyes were fixed unwaveringly upon a distant derrick and the blasted stub of a big drill that hung unmoving above the concrete floor.

But the man's eyes did not consciously record the details of that scene. He saw nothing of the derrick or of the heat waves that made the steel seem writhingly alive; he was looking at something far more distant, something many miles away, something vague and mysterious, hidden miles beneath the surface of the earth.

"Heat," he said at last, as if talking in a dream. "Heat, terrific temperatures — but I can't make it out; I can't see it!"

The younger, broad-shouldered man, whose khaki shirt, thrown open at the neck showed a chest tanned to the black-brown of his face, stopped his restless pacing back and forth in the hot room.

"Yes?" he asked with a touch of irritation in his tone. "There's plenty of heat there — heat enough to melt off the shaft of that high-temp alloy! What the devil's the use of wondering about the heat, Dean? What gets me is this: the shaft has been plugged again. Now, what kind of...."

Dean Rawson's face had not moved a muscle during the other's outburst. His eyes were still fixed on that place that was so far away, yet which he tried to bring close in his mind, close enough to see, to comprehend the mystery that should be so plain.

"Lava wouldn't do it!" he said softly. "No melted stone would melt the Krieger alloy, unless it was under pressure, which this was not. There was no blast coming out of our shaft. Yet we dipped into that gold; we stuck the drill right down into it. But what did we go into the next time? What did we dip into?"

He swung quickly, violently, toward Smithy who was facing him from the middle of the room. He aimed one finger at him as if it were a pistol, and his words cracked out as sharply as if they came from a gun:

"That tube you sent down—that piece of casing! How was it burned? Were there straggling ends, frozen gobs of metal? Did it look like an old-fashioned molasses candy bar that's been melted? Did it?"

"Why, no," said Smithy. "It hadn't dripped any; it was cut off nice and clean."

"Cut!" Rawson almost shouted the word. "You said it, Smithy. So was the shaft of the drill. And if you ever saw a piece of this alloy being melted you know that it's as gummy as a pot of old paint. It was cut, Smithy! Dipping into that melted gold threw us off the track; we were thinking of ramming the drill down into a mess of lava. But we didn't. It was cut off by a blast of flame so much hotter than lava that melted rock would seem cold!"

"And that helps us a lot, doesn't it," asked Smithy, scornfully, "when the flame melts the end of the shaft shut as fast as we open it?"

Dean Rawson's lean, muscular hands took Smithy's broad shoulders and spun the younger man around. "Cheer up," Dean told him. "We've got it licked. Why it doesn't blow out of that shaft like hell out for noon is more than I can see; but the heat's there! We've won!"

"But—" Smithy began. Rawson sent him spinning toward the door in a good-natured showing of strength that his assistant had not yet guessed.

"Soup!" he ordered. "Break out the nitroglycerine, Smithy. Get that Swede, Hanson, on the job; he's a shooter. He knows his stuff. We'll blow open the bottom end of our shaft so it'll never go shut!"

Hanson knew his stuff and did it. But he met Rawson's inquiring eyes with a puzzled shake of his head when the open mouth of the twenty-inch bore gave faint echo of the deep explosion and followed after a time with only a feeble puff of air.

"Like a cannon, she should have gone," Hanson stated. "And she yoost go *phht*!"

"It's open down below," said Rawson briefly. "This is a different kind of a well from the kind you've been shooting."

To the waiting Riley he said: "Hook a bailer onto that cable and send it down. See what you can tell about the hole."

Again ten miles of cable hissed smoothly down the gaping throat. Then it slowed.

"Fifty-two-seven," said Riley, "and she's open. Seven twenty-five! Seven fifty, and we're on bottom!"

"Up," Rawson ordered, "if there's anything left of the bailer. It's probably melted into scrap."

But strangely it was not. It hung from the dangling cable spinning lazily until Riley stepped in to check its motion.

There was a check valve in the bottom—a door that opened inwardly, to take in water and fragments of rock when need arose. Riley, disregarding the possible heat of the twirling bailer, reached for it with bare hands. He drew them back, then held them before him—and a hundred watching eyes saw what had been unseen before: the slow dropping of red liquid from the bailer's end. The same drops were falling from Riley's hands that had touched that end.

"Blood!" The word came from the foreman's throat in one horrified gasp. It ran in a whispering echo from one to another of the watching crew. From far across the hot sands came the rattle of a truck that brought the first of many loads of cement and steel for Rawson's buildings. Its driver was singing lustily:

> "Hark to what I say:
> You're pokin' through the crust of hell
> And braggin' too damn loud of it,
> For, when you get to hell, you'll find
> The devil there to pay!"

But Rawson, looking dazedly into Smithy's eyes, said only: "It's cold—the bailer's cold. There's no heat there."

CHAPTER IV
THE LIGHT IN THE CRATER

Of course it wasn't blood!" said Smithy explosively. "But try to tell the men that. See how far you get. 'Devils!' That's been their talk since yesterday when Riley got smeared up—and now that the bailer's gone we can't prove a thing."

Again he was pacing restlessly back and forth in the little board shack that was Rawson's field head-quarters. Rawson, seated by the window, was looking at tables of comparative melting points. He glanced up sharply.

"You haven't found it yet?" he questioned. "A forty-foot bailer! Now that's a nice easy little thing to mislay."

Riley had followed the excited Smithy into the room; he stood silently by the door until he caught Rawson's questioning glance.

"Forty feet or forty inches," he said, "'tis gone! 'Twas there by the derrick last night, and this marnin'—"

"That's fine," Rawson interrupted with heavy sarcasm. "I haven't enough down below ground to keep my mind occupied—I need a few mysteries up top. Now do you really expect me to believe that a thing like that bailer has been carried off?"

This time it was Smithy who interrupted. "You can just practise believing on that, Dean," he said. "When you get so you can believe a forty-foot bailer can vanish into thin air, then you'll be ready for what I've got. This is what I came in to tell you: that one truckload of steel grillage beams for the turbine footings—they were put out where we surveyed for the first power house—dumped on the sand...."

"Well?" questioned Rawson, as Smithy paused. His look was daring Smithy to say what he knew was coming.

"Five tons of steel beams," said Smithy softly, "gone—just like that! Just a hollow in the sand!"

The big figure of the Irish foreman was still beside the door. Rawson saw one clumsy hand make the sign of the Cross; then Riley held that hand before him and stared at it in horror. "Divils' blood," he whispered. "And I dipped my hands in it. Saints protect us all!"

"That will be all of that!" Dean Rawson's usually quiet voice was as full of crackling emphasis as if it had been charged with electrical energy. "If anyone thinks that I have gone this far, just to be scared out by some dirty sabotage...."

"I see it all. I don't know how they did it, but it's all come since the gold was found. Someone else wants it. They think they can scare off the men, maybe take a pot-shot at me, come back here and clean up later on, pull up gold by the pailful, I suppose—"

Riley leaped forward and banged his big fist down on the table. "Right ye are!" he shouted, until loitering men in the open "street" outside stared curiously. "Divils they are, but they're the kind of divils we know how to handle. And now I'll tell ye somethin' else, sir: I know where they are hidin'.

"There was no work for anyone last night, but I'm used to bein' up. I couldn't sleep. I was wanderin' around, thinkin' of nothin' at all out of the way, and I thought I saw some shadows, like it might be men, way off on the sand. Then later over to the old ghost town, d'ye mind! I saw a light, a queer, green sort of light. Sure, a fool I was callin' meself at the time, but now I believe it."

Dean Rawson had crossed the room while the man was still speaking. He dragged a wooden case from beneath his cot and smashed at the lid with a wrecking bar. Then he reached inside and drew forth a blue-black .45.

He tossed the pistol to Riley. "Know how to use one of these?" he asked. The manner in which the big Irishman snapped open the side ejection was sufficient answer. Dean handed another gun to Smithy, then pulled out more and laid them on his cot together with a little pile of cartridge boxes.

"You're all right, Riley," he said. "Just keep your head. Don't let your damned superstitions run away with you, and I wouldn't ask for a better man to stand alongside of in a scrap."

The foreman beamed with pleasure: Rawson went on in crisp sentences:

"Take these guns. Take plenty of ammunition. Pick five or six men you know you can depend on. Mount guard around this camp to-night. I'll post an order saying you're in charge—and I'm telling you now to use those guns on anything you see.

"Smithy," he said to the other man who had been quietly listening, "you and I are going to start for town. Only Riley will know that we're gone for the night. We'll have a little listening post of our own up here in the hills."

But Rawson postponed their going. More material was arriving; one casting in particular needed all the men and Rawson's supervision to place it on the sand where an erection crew could swing it into place at some later date. And then, when he and Smithy had driven away from camp with the distant city as their announced destination, Rawson still did not go directly to the mountain grade. He swung off instead where rolling sand-hills blocked all view from the camp, and he headed the car into a gusty wind that brought whirling clouds of dust; they almost obscured the crumbling walls at the volcano's base.

The ghost towns that are found here and there in the forsaken wilderness of the West are depressing to one who walks their empty streets. Little Rhyolite was no exception. In gray, ghostly walls, empty windows stared steadily, disconcertingly like sockets of dead eyes in tattered, weatherbeaten skulls.

Dean and Smithy walked among the roofless ruins. Lizards, the color of the cold, gray walls, slipped from sight on silent, clinging feet. Once a sidewinder, almost invisible against the sand, looped away from the intruders with smooth deliberation.

"No marks here," said Rawson at last. "Even an Indian can't read sign in this ashy sand when the wind has dusted it off."

He turned his head from a whirl of fine ash where the wind, sweeping around a wall of stone, was scouring at a sand dune's sloping side.

"Dean," said Smithy, "old Riley may have been looking for banshees when he saw these lights. Superstitious old cuss, Riley! Maybe there wasn't anything here. But, Dean, there's some confoundedly funny things happening around here."

"Are you telling me?" Rawson asked grimly. "But we want to remember one thing," he added: "We've punched a hole in the ground, and we've got into a place that is hot enough to melt Krieger alloy one minute and is stone cold the next. That's disturbing enough, but we don't want to get that mixed up with what's happening up top. There's dirty work going on—"

He stopped. His eyes, that had never ceased to search for some mark of special meaning, had come to rest upon an object half hidden in the sand. He stooped and picked it up.

"Now what the devil is this?" Smithy began. But Rawson was staring at the smooth lava block that was in his hand. It was tapered; it was pierced through with a straight, smooth hole, and its base was round and ringed as if it had been held in a clamp.

"That," he said at last, "was brought in from outside. Outside, Smithy— get that."

Dean Rawson's face was wreathed in a sudden smile of pure pleasure. "No, I don't know what the darn thing is," he admitted. "And I don't care. But I know that someone, or some bunch of someones—outsiders—are trying to horn in. I might even go so far as to say that I suspect the power monopoly gentlemen. I think they have started in on us, plan to run off our men, interfere in every way and drive me out of the field with the boring a failure. Smithy, I begin to think I'm going to enjoy this job!"

Again the hot wind, only beginning to cool with the setting of the sun, swept around the building where they stood and tore at the hill of sand. "Come on," said Rawson. "It's getting dark. We'll get up to our lookout—"

"Hold on!" called Smithy sharply.

Rawson turned. Smithy was rubbing his eyes when the whirl of wind-borne sand had passed; he was staring at the sand dunes.

"I'm seeing things, I guess," he said. "I thought for a minute there was a hole there, and the sand was slipping. I'm getting as bad as Riley."

The two went back through the gathering shadows to their waiting car. And Smithy's involuntary shiver told Rawson that he was not the only one to feel a sense of relief at the sound of the exhaust as their car took them away from the dead bones of a dead city in a barren, trackless waste.

The shoulder of rock, where the mountain road swung out, gave a comprehensive view of camp and desert and the encircling mountains. Above in a vault of black was the dazzling array of stars as the desert lands know them; so low they were, the ragged, broken tops of the three ancient craters seemed touching the warm velvet of the sky on which the stars were hung. Beyond their smooth slopes a spreading glow gave promise of the rising moon.

Rawson headed the car downgrade in readiness for a quick return; he ran it close to the inner wall of rock out of which the road had been carved, then seated himself on the outer rim without thought of the thousand-foot sheer drop beneath his dangling legs. With a glass he was sweeping the foreground where the scattered lights of the camp were like vagrant reflections of the stars thrown back to them from the dead sea of sand.

"Riley's on the job," he told Smithy when he passed over the glass later on. "And I've got my pocket portable." He took the little radio receiver from his pocket as he spoke. "Riley will signal me from my office if he sees anything."

The moon had cleared the mountains; its flood of light poured across their rugged heights and filled the bowl of Tonah Basin as some master of a great theatrical switchboard might have flooded a dark stage with magic illumination, half concealing, transforming whatever things it touched.

All the hard brilliance of sunlit sands was gone. The rolling dunes were softly mellow; the more distant mountains were dream-peaks. Half real, they seemed, and half imagined in a veil of haze. Even the buildings, the scattered piles of material, the gaunt skeleton of the derrick—their stark blackness of outline and clear-cut shadow were gone; the whole land was drenched in the mystery and magic of a desert moon.

Rawson and the man beside him were silent. Even a mind perplexed by unanswerable problems must pause before the witchery of nature's softer moods.

"If Riley were here," said Smithy softly at last, "he wouldn't be seeing any devils. Fairies, pixies, the 'little people'—he'd be seeing them dancing."

Rawson shot his companion a sidelong, appraising glance. He had never penetrated before to this sub-stratum of Smithy's nature. He had never, in fact, felt that he knew much about Smithy, whose past was still the one topic that was never mentioned. He saw his thick mop of black hair and the profile of his face as Smithy stared fixedly down toward the sleeping camp. It was a matter of a minute or so before he knew that the head was outlined against an aura of red light.

Smithy was seated at his right. Off beyond him the three extinct craters made a dark background where the moonlight had not yet reached to their inner slopes. Smithy's head was directly in line with the largest crater's irregularly broken top; and about it was the faintest tinge of red.

For a moment the light flamed close; it seemed to be hovering about the head of the silent, seated man. Then Rawson moved, looked past, and found a true perspective for the phenomenon. One rugged cleft in the rim of the crater's cup made a peephole for seeing within. It was plainly red—the light came from inside the age-old throat.

"It's alive!" Rawson whispered in quick consternation. Almost he expected to see billowing clouds of smoke, the fearful pyrotechnics of volcanic eruption.

He sensed more than saw that Smithy had not turned his head. "Look!" he was shouting by now. "Wake up, Smithy! Good Lord!"

He stopped, open-mouthed. The red glow had meant volcanic fires; to have it change abruptly to a green radiance was disconcerting.

Green—pale green. Only through the gap, like a space where a tooth was missing in the giant jaw, could Dean Rawson see the changed light. Only from this one point could the view be had—there would be nothing visible from the camp below. And as quickly as it had come all thought of volcanic fires left him; he knew with quick certainty that this was something that concerned him, that threatened, and that was linked up with the other threatening, mysterious happenings of the recent nights and days.

Still Smithy had not turned. Rawson felt one quick flash of annoyance at his helper's dullness — or indifference; then he knew that Smithy's dark-haired head was reached forward, that he was bending at a precarious angle to stare below him into the valley. Then:

"They're there!" said Smithy in a hushed voice, as if someone or something on that desert floor far below might hear and take alarm. "Look, Dean. Where's your glass? What are they?"

His cautious whispering was unnecessary. Below them a thin line of light pierced the darkness; another; then three more in quick succession before the sharp crack of pistol fire came to the men a thousand feet above. Rawson had snatched up his binoculars.

"To the left," Smithy was directing. "Off there, by the big casting. Great Scott! what's that light?"

Rawson got it in the glass — a single flash of green that cut the blackness with an almost audible hiss. It was gone in an instant while a man's voice screamed once in fear and agony, one scream that broke like brittle steel in the same instant that it began.

Dean found the big casting in the circle of his glass. There were black figures moving near it; they were indistinct. He changed the focus — they were gone before he could get their images sharp.

But the casting! Plainly he saw its great bulk that many men had worked to ease down to the sand. It was outlined clearly now until its edge became a blur, until the sand rolled in upon it, and its black mass became a circle that shrank and shrank and vanished utterly at the last.

"It's gone!" Rawson shouted. "It sank into the sand! I saw it...."

He was running for the car. A clamor of voices was coming from below; the sound died under the thunder of the car's exhaust as Rawson gave it the gun and sent the big machine leaping toward the waiting curves.

CHAPTER V
THE ATTACK

Every light of the camp was on as Rawson and his assistant approached. A shallow depression in the sand marked the place where the big casting had been. Beyond it a hundred feet was a black swarm of men that parted as the car drew near. They had been gathered about a figure upon the sand.

Dean sensed something peculiar about that figure as the big car ploughed to a stop. He leaped out and ran forward.

He knew it was Riley there on the ground, knew it while still he was a score of feet away. Only when he was close, however, did he realize that the body ended in two stubs of legs; only when he leaned above him did he know that the Irish foreman's big frame had been cut in two as if by a knife.

The severed legs lay a short distance beyond the body; they had fallen side by side in horrible awkwardness, their stumps of flesh protruding from charred clothing — and suddenly, shockingly, Rawson knew that the flesh of body and legs had been seared. The knife had been hot — its blade had been forged of flame!

He heard Smithy cursing softly, unconsciously, at his side.

"The green light," Smithy was saying in horrified understanding. "But who did it? How did they do it? Where did they go?"

"Quiet!" ordered Rawson sharply. He dropped to his knees beside the mutilated body. Riley's eyes had opened in a sudden movement of consciousness.

The voice that came from his lips was a ghastly whisper at first, but in that stricken thing that had been the body of Riley, foreman of the night drilling crew, some reservoir of strength must still have remained untapped.

He drew upon it now. His voice roared again as it had done so many times before through the Tonah Basin camp. It reached to every listening ear where crowding men stood hushed and motionless; and the overtone of terror that altered its customary timber was apparent to all.

"Devils!" said Riley. "Devils, straight out o' hell!... I saw 'em — I saw 'em plain!... I shot — as if hot lead could harm the imps of Satan....

"Oh, sir," —his eyes had found those of Dean Rawson who was leaning above—"for the love of hivin, Mister Rawson, do ye be quittin' drillin'. The place is damned. L'ave it, sir; go away...."

His eyes closed. But he started up once more; he raised his head from the sand with one final convulsive movement, and his voice was high and shrill.

"The fire! The fire of hell! He's turnin' it on me! God help...."

But Riley, before his failing mind could recall again that torturing jet of flame, must have slipped away into a darkness as softly enveloping as the velvet shadow world behind the low-hung stars. Rawson's hand that felt for a moment above the heart, confirmed the message of the closed eyes and the head that fell inertly back.

He came slowly to his feet.

"Keep the floods on!" he ordered. "Take command of the armed guard, Smithy; keep the whole camp patrolled."

Then to the men:

"Boys, Riley was wrong. He believed what he said, all right, but Smith and I know better. Don't worry about devils. These're just some dirty, skulking dogs who got away with murder this time but who won't do it again. We know where they're hiding. I'm checking up on them right now. After that you'll all get a chance to square accounts for poor old Riley!"

But the casting!" Smithy protested when he and Rawson were alone. "You can't explain that disappearance so easy, Dean."

"No, I can't explain that," Rawson's words came slowly. "They've got something that we don't understand as yet—but I'm going to know the answer, and I'm going to find out to-night!"

He was seated behind the wheel of his old car.

"I'm as good a desert man as there is in this crowd," he told Smith. "And it's my fight, you know. I'm going alone. But there'll be no fighting this trip; I'll just be scouting around."

He leaned from the car to grip Smithy's shoulder with a hand firm and steady.

"You didn't see the crater when the show was on. You think that I'm crazy to believe it, but up in that crater is where I'll find the answer to a lot of questions. Lord knows what that answer will be. I've quit trying to guess. I'm just going up there to find out."

He was gone, the rear wheels of the car throwing a spray of sand as he started heedless of Smithy's protests against the plan. Rawson was in no mood to argue. He must climb the mountain while it was night; under the sun he would never reach the top alive. He would go alone and unseen.

He swung wide of the deserted town at the mountain's base. The spectral walls of Little Rhyolite still showed their empty windows that stared like dead eyes, and the man guided his car without lights along a hidden stretch of hard, salt-crusted desert. He felt certain that other eyes were watching.

He began his climb at a point five miles away. The slopes that seemed smooth and hard from a distance became, at closer range, a place of wind-heaped, sandy ash, carved and scoured into fantastic forms. But its very roughness offered protection, and Rawson fought the dragging sand, and the gray, choking ash that dried his throat and cut it like emery, without fear of being observed.

He fought against time, too. Above Little Rhyolite, whatever mysterious men were making the ascent would find the going easy. There were windswept areas, long fields of pumice; a man could make good time there. Rawson had none of these to aid him. He cast anxious glances toward the eastern sky as he struggled on, till he saw gray light change to rose and gold—but he stood in the titanic cleft in the crater's rim as the first straight rays of the sun struck across.

The volcano's top had been stripped clean by the winds of countless years. Rocks, black, brown, even blood-red, were naked to the pitiless glare of the sun. Their colors were mingled in a weird fantasy of twisted lines that told of the inferno of heat in which they had been formed.

They towered high above the head of Dean Rawson as he stood, panting and trembling with exhaustion. The cleft before him had become enormous: it was a canyon, half filled with pumice and coarse ash.

Rawson stood for long minutes in quiet listening. At the canyon's end would lie the crater, and in that crater he would find.... But there was no slightest picture in his mind of what he might see. He knew only that he himself must remain unseen. He went forward cautiously.

Rocky walls; a floor of sand where his feet left no mark. He was watching ahead and above him. His gun was ready in his hand; he did not propose to be ambushed. He moved with never a sound.

The silence persisted; no living thing other than himself lent any flicker of motion to the scene. Not even a lizard could hope for existence amid these dead and barren heights. He was alone—the certainty of it had driven deeply into his mind before the canyon end was reached. And, desert man though he was and accustomed to traveling the waste places of the earth, Rawson learned a new meaning and depth of solitude.

Here was no voiceless companionship of trees or brush or cactus; no little living things scuttled across the rocks—he was alone, the only speck of life in a place where life seemed forbidden.

So sure of this was he that he stepped boldly from the canyon's end. He knew before he looked that he would see only more of the same desolation. And his mind was filled equally with anger and disappointment.

Something was opposing him! Something had come into their camp—had killed old Riley. And he, Rawson, had been so sure he would find traces here that would allow him to give that opposing force a name....

He stared out from the rocky cleft into a sun-blasted pit. Already the rising sun was pouring its energy ever the jagged rim of bleak rocks and down into the vast throat, choked and filled with ash.

It sloped gently from all sides, the gray-brown powder that had been coughed from within the earth. It made a floor where Rawson could have walked with safety. But he did not go on.

"Damn it!" he said with sudden savagery. "What a fool I was to think of finding anyone here. Who would ever pick out a spot like this for a base of operations?"

He stared angrily at the floor of ash, at the black, outcropping masses of tufa. He was angry with himself, angry and baffled and tired from his climb. Far down in the vast, shallow pit blazing sunlight glinted from massive blocks whose sides were mirror-smooth. A whirl of wind eddied there for a moment and lifted the dust into a vertical gray column—the only sign of motion in the whole desolate scene. Rawson turned and tramped back toward the long hot descent to the floor of the Basin.

He tried to maintain an air of confidence before the men. He kept them busy placing and stacking materials; to all appearances the work would go on despite the mysterious happenings of the night.

Dean even prepared to resume drilling operations. He sent down another bailer on the end of the ten-mile cable, but he left it there; he did not care to raise it and risk more inexplicable results with the consequent destruction of the men's morale.

"Too late to do any more," he said to Smithy that afternoon. "We'll drop all work—let the men get a good night's sleep. I'll take guard duty to-night, and you can run the job to-morrow."

There were men of the drilling crew standing near, though Rawson was handling the hoisting drums himself. A ratchet release lever hooked its end under a ring on Rawson's hand and pinched the flesh. Dean made this an excuse for waiting a moment while the drillers walked away.

"Ought not to wear it, I suppose," he said, and dabbed at a spot of blood under the gold band. "But it's an old cameo—it belonged to my Dad."

He was showing the ring to Smithy as the men passed from hearing.

"Don't want to be seen talking," he explained tersely. "Mustn't let the men know we are on edge—they're about ready to bolt. But you be ready for a call. Have your men armed. I am looking for more trouble to-night."

The two were laughing loudly as they followed the men toward the building where the cook was banging on an iron tire that served as a bell.

Some three hours later Rawson was not smiling as he climbed the steel ladder of the great derrick; he was grimly intent upon the job at hand.

All thought of his drilling operations had gone from him. He was not anxious about the project. This was merely an interruption; the work would go on later. But right now there was an enemy to be met and a mystery to be solved.

A rifle slung from his shoulder bumped against him satisfyingly as he climbed. A man was on duty at a master switch—he would flood the camp with light at the rifle's first crack.

Dean seated himself at the top of the derrick. The cylinder of a huge floodlight was beside him. Beyond was the massive sheave block; the cables ran dizzily down to the concrete drilling floor so far below. And on every side the quiet camp spread out dark and silent in the night. Dean surveyed it all with satisfaction. Nothing would get by him now.

But his further reflections were not so satisfying.

"Who did it? How? Where did they go?" He was echoing Smithy's questions and finding no ready answers. And that flame-thrower that had cut down old Riley—how was that worked? Its one green flash had been almost instantaneous.

He was puzzling over such futile questioning when he saw the first sign of attack.

At the foot of the derrick was the hoisting shed. Except for that, there was clear sand for a radius of fifty feet around the derrick's base. Dean was staring suspiciously at that open space almost directly underneath.

Moving sand! He hardly knew what he had seen at first. Then the sand at one point bulged upward unmistakably.

For one instant Dean's thoughts shot off at a tangent. It was like the work of a huge gopher—he had seen the little animals break through like that. Then the sand parted, and something, indistinct, blurred, dark against the yellow background, broke from cover.

Rawson swung the rifle's muzzle over and down. Below him the vague shadow had moved. Dean caught the blurred mass beyond his sights, then swung the weapon aside. Who was it? He would have a look first.

The thin crack of his rifle ripped the silence of the sleeping camp. Dean had aimed to one side and he regretted it in the instant of firing. For, in the same second, there had come from the moving shadow the gleam of starlight reflected upward from polished metal.

Dean swung the rifle back. He fired quickly a second time. Beside him the big light hissed into action and the whole camp sprang to sudden, blazing light. And through the quick brilliance, more dazzling even than the white glare itself, was one blinding line of green flame.

Dean saw it as it began. It came from the dim shadow that had sprung suddenly into sharp outline as the big lights came on. He saw the figure. He sensed that it was a man, though he knew vaguely that the figure was grotesque and hideous in some manner he had no time to discern.

The thin line of green flame ripped straight out, swinging in a quick, sweeping trajectory, slashing through the steelwork of the great derrick itself!

Dean knew he was lost in the blinding instant while that fiery jet was sweeping in a fan-shaped sector of vivid green. A knife of flame! It had destroyed a man: it was now cutting down a framework of steel as well!

The derrick was falling as he fired again. There came a crushing jar downward as the metal melted and failed, and the wild outward swing in the beginning of the toppling fall. In the mind of Dean Rawson was but one thought: the sights—and a something blurred beyond—a trigger to be pressed.

He was still firing when the shriek of torn steel went to thundering silence, and even the lights of Tonah Basin Camp were swallowed up in the whirling night....

CHAPTER VI
INTO THE CRATER

Smithy's agonized face was above him when he came back to life. "God!" Smithy was breathing. "I thought you were gone, Dean! I thought you were dead!"

As it had been with Riley, there was one thought uppermost in Rawson's bewildered mind: "The fire!" he choked. "He's swinging it...."

Then, after a time: "The derrick—it's falling! I went down with it!... I hit—"

"I'll say you did," said the relieved Smithy. "The derrick smashed across the bunkhouse, snapped you off, sent you skidding down the side of a sand dune. It darned near scoured the clothes off you at that."

Slowly Rawson began to feel the return flow of life through his body; the shock had jarred every nerve to insensibility. Slowly he remembered and comprehended what had happened.

He was in his little office; he recognized his surroundings now. The windows were open. Outside the sun was shining. He realized at last the utter silence of that outer world.

He tried to raise himself from the cot, but fell back as his surroundings began to spin. "The camp!" he gasped weakly. "The men—I don't hear them."

"Gone!" Smith told him, while his eyes narrowed at some recollection and his hand came up unconsciously to a bruise of his cheek. "They beat it—went last night after the derrick fell. I tried to stop them. The fools were crazy with fear—devils, hell, all that kind of stuff. It all wound up in a fight—I couldn't hold 'em.

"You've got to get better kind of fast," he told Rawson. "We've got to get out of here ourselves—that flame-throwing stuff is too strong for me to take."

Rawson suddenly remembered the vague figure that had directed that flame. "Did I get him?" he demanded eagerly.

"You got him, yes, but then a whole swarm of things boiled up out of nowhere and carried him off! We weren't any of us close enough to see. The men said they were devils; I'm not sure they were wrong, either. Dean, old man, we're up against something rotten. We've got to get fixed for a fight; we can't handle this by ourselves."

Rawson was silent. He spoke slowly at last:

"You mean we've got to quit—quit without knowing what we're up against. Can you imagine what they'll say to me back in town? Scared out, licked by something I've never even seen!"

"Scared?" Smithy inquired. "You couldn't find a better word for it if you hunted through the whole dictionary. Scared? Why, say, I'm so damn scared I'm shaking yet, and the only thing that will cure me of it is to look at those devils along the top of a machine gun! We'll go catch us some equipment and a few service men—"

"You're a good guy, Smithy," Rawson reached out and gripped one brown hand. "And we'll do as you say; but first I've got to get a line on things. I'm becoming as irrational as the men. I'm imagining all sort of crazy things."

"You don't have to imagine them." Smithy's voice was strained; it showed the tension under which he was laboring. "Men or beasts—God knows what they are!—but when they come up from nowhere—"

"Out of the sand," Rawson explained.

Smithy stared at him. "Out of the sand," he repeated. "Then, when they cut a man in two, melt steel as if it were butter, pull a few tons of metal down out of sight as easy as we would sink it in the ocean, flash their lights over in the ghost town, up on top of a volcano—"

"Stop!" shouted Rawson unexpectedly. Some sudden gleam of understanding had flashed through his mind. He dragged himself to his feet and staggered to the doorway where he clung until the nausea of a whirling world had passed. "The dust! The dust!" he gasped.

Smithy put a hand on his shoulder. Plainly he thought Rawson out of his mind. "Easy, old-timer," he cautioned. "We'll get out of here. I hate to make you walk in the shape you're in, but the dirty cowards ran off with the trucks. They even took your car; there isn't a thing here on wheels."

But Rawson did not hear. He was staring off across the sand, and he was muttering bitter words.

"Fool! Oh, you utter fool!" he said. "The dust—the dust." Then he let the roughly tender hands of Smithy guide him back to the cot where he fell into a troubled sleep.

The comparative coolness of dusk was tempering the feverish midday heat when Rawson awoke. And, strangely, his troubles and all his conflicting plans had been simplified by the magic of sleep. His course was entirely plain. He was going to the crater again.

"What's there?" Smithy demanded. "What do you think that you'll find?"

"I don't know," was the reply.

"Then why—what the devil's the idea?"

"It's my job. They put it up to me, Erickson and his crowd. I've got to go."

And nothing Smithy could say seemed able to reach Rawson and swerve him from his single idea.

"You'll be safe on the road," Rawson told him, while he filled a canteen with water in preparation for his own trip. "You can get to the highway by morning."

Smithy did not trouble to reply. Was Rawson out of his mind? He could not be sure. Certainly he had got an awful bump, but there were no bones broken. However, it might be that he was still dazed—a crack on the head might have done it.

But there was no use in further argument, he admitted to himself. Dean was going to the crater again—there was no stopping him—but he was not going alone; Smithy could see to that.

Again Rawson took the more difficult ascent. They went first to the ghost town: the slope above Little Rhyolite would save weary miles. But, once there, they knew that the route was not a place where they would care to be in the night. The realization came when Smithy, walking where they had been the day before, passing the sand dune where the wind had been scouring, seized Rawson's arm.

"I thought so," he said softly. "I thought I saw something there the other day, but the sand fell in and hid it. I didn't know the old-timers went in for subways in Little Rhyolite."

And Rawson looked as did Smithy, in wondering amazement, at the roughly round opening in the sand, a tunnel mouth, driven through the shifting sands—a tunnel, if Rawson was any judge, lined with brown glistening glass.

Understanding came quickly.

"The jet of flame!" he exclaimed half under his breath. "They melted their way through; the sand turned to glass; they held it some way for an instant while it hardened." He walked cautiously toward the dark entrance and peered inside.

Darkness but for the nearer glinting reflections from walls that had once been molten and dripping. The tunnel dipped down at a slight angle, then straightened off horizontally. Rawson could have stood upright in it with easily another two feet of headroom to spare.

"And that," said Smithy, "is how the dirty rats got over to the camp. Like moles in their runway. No wonder they could pop up from nowhere. But, Dean, old man, I'm thinkin' we're up against something we haven't dared speak of to each other. Don't tell me that it's just men we've got to meet—"

"Wait," Rawson begged in a hushed whisper. "Wait till we know. That's why I didn't dare go out without something definite to report. We'll go up—but not here. We'll get a line on this up top."

He led the way from the crumbling walls and skirted the mountain's base to the place where he had climbed before. And, with the help of a supporting arm at times, he found himself again in the great cleft in the rocks.

Darkness now made the passageway a place of somber shadows. The broad cupped crater lay beyond in silent waiting; the vast sand-filled pit seemed, under the starlight, to have been only that instant cooled. The twisted rocks that formed the rim had been caught in the very instant of their tortures and frozen to deep silence and eternal death: the black masses of tufa, protruding from the packed ashy sand might have been buried by the smothering mass but a moment before. It was a place of death, a place where nothing moved—until again the breeze that whirled gustily over the saw-tooth crags snatched at the sand in that lowest pit and drew it up in a spiral of dust.

The word was on Rawson's lips. "Dust—dust in the crater. Fool! I said I could read sign; I thought I was a desert man."

"Dust? And why shouldn't there be dust? How do you usually have your volcanoes arranged, old man?"

"Fine dust!" Rawson interrupted in the same whisper. He was glancing sharply about him as if in fear of being overheard. "See, the wind is blowing it. Coarse sand and pumice—that's to be expected; but light dust in a place that the winds have been sweeping for the last million years! I don't have them arranged that way, Smithy—not unless the sand has been recently disturbed!"

He moved soundlessly across the sand. There was no chance for concealment; the surface was too smooth for that. Yet he wished, as he moved onward down the long, gentle slope, that he had been able to keep under cover. In all the wide bowl of the great crater top was nothing but dead ashes of fires gone long centuries before, coarse, igneous rock—nothing to set the little nerves of one's spine to tingling. Rawson tried to tell himself he was alone. Even the gun in his hand seemed an absurd precaution. Yet he knew, with a certainty that went beyond mere seeing, that invisible eyes were upon him.

The blocks were massive when he drew near to them. They were buried in the sand, their sides like mirrors, their edges true and straight. "Crystals," Rawson tried to tell himself, but he knew they were not.

Gun in hand, he moved among the great rocks. Open sand lay beyond, running off at a steeper pitch to make a throat—a smaller pit in the great pit of the crater itself. Rawson noted it, then forgot it as he stooped for something that lay half hidden, its protruding end shining under the light of the stars, as he had seen it gleam before at the derrick's base.

He snatched up the metal tube, noting the lava tip, and that it was like the one Smithy had found in the ghost town. The tube, clearly, was part of some other mechanism, and Rawson realized with startling suddenness that he was holding in his hand the jet of a flame-thrower—the same one, perhaps, that had almost sent him to his death.

The thought, while he was still thinking it, was blotted from his mind. He was thrown suddenly to the sandy earth; the sand was slipping swiftly from beneath his feet; he was scrambling on all fours, clawing wildly for some anchorage that would keep him from being swept away.

He touched a corner of shining stone, drew himself to it, reached its slanting side, then scrambled frenziedly to the top and threw himself about to face the place of slipping sands. But where the sand had been, his wildly glaring eyes found only a black hole—a vertical bore, like the ancient throat of the volcano; and this, like the tunnel in the sand, was lined with smooth and glistening glass.

It was black at first, a yawning, ominous maw, till the polished sides caught a reflection from below and blazed red with the glare of hidden fires.

No time was needed for Dean's quick searching eyes to grasp the meaning of the change. Whatever had menaced the camp had set this trap. He swung sharply to leap from the block, but stopped at the sight of Smith's chunky figure coming slowly across the sand.

"Back!" he shouted. His voice was almost a scream, shrill and crackling with excitement. "Get back, Smithy! I'm coming!"

He would have leaped. Below the block the sand bulged upward as a yellow animal-thing came clawing up into the night. Dimly he saw it—saw this one and the others that must have been hidden in the sand. They were between him and Smithy! A blaze of red came from behind him—there must be others there! He snatched his gun from its holster as he turned.

Flames were hissing into the darkness, five or six of them in lines of hot crimson fire. They changed to green as he watched, and the livid light spread out in ghastly illumination over the creatures that directed them.

He saw them now—saw them in one age-long instant while he stood in horror on the black shining rock. He saw their heads, red-skinned, pointed, their staring eyes as large as saucers—owl-eyes. They were naked, and their bodies, that would have been almost crimson in the light of day, were blotched and ghastly in the green light. And each one held in long clawlike hands a thing of shining metal—a lava tip like the one he had found projected and ended in the hissing line of green.

A flame slashed downward. For one sickening second he waited to feel the heat of it, though it was many feet away; in his mind he cringed involuntarily from the ripping knife-cut of the fiery blade that would blast the life from him; then he knew that the flame had passed—it was tearing at the rock beneath his feet. And the cold stone turned to liquid fire at that touch.

It leaped in a splashing fountain to the sand. The blaze turned the whole pit to flame. On even the farthest rugged crag of the crater's rim the red light glowed. Before Rawson could raise his own weapon the blast had torn the rock from beneath his feet. The great mass tipped, rolled. Rawson's arms were flung wide in an effort to save himself. Then below him was the black throat with its walls of glass: he was plunging headlong into it, turning as he fell—and somewhere, far down in that throat, was the red glow of waiting fires. He saw it again and again as he fell....

CHAPTER VII
THE RING

"Smithy," Rawson had called him when he found the youngster fighting gamely with death in the heat of Tonah Basin. And Gordon Smith was the name on the company records. Yet he remained always "Smithy" to Rawson, and the name, which Rawson never ceased to believe was assumed, became a mark of the affection which can spring up between man and man.

Town after town is fired by the emerging Red Ones as Rawson lies helpless, a prisoner, far down in their home within the earth.

And now Smithy stood like a rigid carven statue in the midst of a barren sandy waste in the vast cup of a towering volcano top—sand that was in reality coarse pumice and ash. This was a place of death, a place where raging fires had left nothing for plant or animal life. And, over all, the desert stars shone down coldly and added to the desolation with their own pale light.

Smithy had seen Rawson pull himself to the top of the great square-edged rock. Sensing that danger of some sort was threatening, he had started to run to the aid of the struggling man. Then came Rawson's cry.

"Back!" he shouted. "Get back, Smithy! I'm coming—"

But he did not come; and Smithy, halted by the command, was frozen to sudden, panic-stricken immobility by that which followed.

He saw the leaping things, like grotesque yellow giants. They came from the sand; then red ones leaped up from the open throat that had suddenly formed. They held flame throwers, the red ones; and the green lines of fire melted the rock from beneath Rawson's feet. All in the one second's time, it was done, and Rawson's body, his arms wide flung, was hurtling downward into the waiting throat and the threatening red glow from within. Then the carriers of the flame throwers vanished again into the pit, and there was left only a huddle of giant figures that tore at the loose sand and ash with their hands.

They threw the material in a continuous stream; the air was full of cascading sand. To Smithy they were suddenly inhuman—they were almost animals; men like moles. And they and their companions had captured Dean Rawson—sent him to his death. Slowly the watching man raised himself from the crouched position that had kept him hidden.

They were through with their work, these great yellow-skinned naked men—or mole-men. Six of them—Smithy counted them slowly before he took aim—and two were armed with flame-throwers.

Smithy rested his arm across the little hummock of gritty ash that had sheltered him and sent six flashes of flame through the night toward the cluster of bodies.

He made no attempt to aim at each individual—the shapes were too shadowy for that. And he had no knowledge of what other weapons they might have. One thing was sure: he must take no chances on facing the red ones single-handed. He rammed his empty pistol back into its holster as he turned and ran—ran with every ounce of energy he possessed to drive his flying feet across the crater floor, out through the cleft in the rocks and down the steep mountainside.

He was stunned by the suddenness of the catastrophe that had overtaken them. The horror of Dean Rawson's going; the fearful reality of those "devils from hell" that old Riley had seen—it was all too staggering, too numbing, for easy acceptance. Time was required for the truth to sink in; and through the balance of the night Smithy had plenty of time to think.

He dared not go back to the camp where ripping flashes of green light told him the enemy was at work. And then, even had it been possible to creep up on them in the darkness, that one chance vanished as the desert about the camp sprang into view. One after another the buildings burst into flame, and Smithy was thankful for the concealment of the vast, empty desert.

The embers were still glowing when he dared go near. This enemy, it seemed, worked only at night, and Smithy waited only for the sun to show above distant purple ranges. It had been their enemy once, that fiercely hot sun; they had fought against the heat—but never had the sun wrought such destruction as this.

Smithy looked from haggard, hopeless eyes upon the wreckage of Rawson's camp. For the men who had worked there, this had meant only a job; to Smithy it had been a fight against the desert which had defeated him once. But to Rawson it meant the fruit of years of effort, the goal of his dreams brought almost within his reach.

Smithy looked at the smoldering heaps of gray where an idle wind puffed playfully at fluffy ash or fanned a bed of coals to flame. Twisted steel of the wrecked derrick was still further distorted; the enemy had ripped it to pieces with his stabbing flames. Even the unused materials, the steel and cement that had been neatly stacked for future use — the flames had been turned on it all.

And Smithy, though his voice broke almost boyishly from his repressed emotion, spoke aloud in solemn promise:

"It's too late to help you, Dean. I'll go back to town, report to the men who were back of you, and then.... They're going to pay, Dean! Whoever — whatever — they are, they're going to pay!"

He turned away toward the mountains and the ribbon of road that wound off toward the canyon. Then, at some recollection, he swung back.

"The cable's still down — he would have wanted it left all shipshape," he whispered.

Where the derrick had stood was the mouth of the twenty-inch casing. The cable that ran from it was entangled with the wreckage of the derrick, but it had not been cut. Smithy set doggedly to work.

A little gin-pole and light tackle allowed him to erect a heavier tripod of steel beams; it hoisted the big sheave block into place, and gave Smithy's two hands the strength of twenty to rig a temporary hoist. The juice was still on the main feed line, and the hoisting motors hummed at his touch. The ten miles of cable wound slowly onto the drums.

"It's nonsense, I suppose," he told himself silently. But something drove him to do this last thing — to leave it all as Rawson would have had it.

The long bailer came out at last; there was just room to hoist it clear and let it drop back upon the drilling floor. A glint of gold flashed in the sunlight as Smithy let the long metal tube down, and he broke into voluble cursing at sight of the bit of metal that was caught near the bailer's top.

The gold had started it all! That first finding of the gold on the big drill had begun it.... He crossed swiftly to the gleaming thing that seemed somehow to symbolize his loss.

He stooped to reach for it, intending to throw it as far as he could. Instead he stood in an awkward stooping attitude — stood so while the long uncounted minutes passed....

His eyes that stared and stared in disbelief seemed suddenly to have turned traitor. They were telling him that they saw a ring — a cameo — jammed solidly into the shackle at the bailer's end. And that ring, when last he had seen it, had been on Dean Rawson's hand! Dean had caught it; he had hooked it over a lever in this very place — and now, from ten miles down inside the solid earth, it had returned. It meant — it meant....

But the stocky, broad-shouldered youngster known as Smithy dared not think what it meant. Nor had he time to follow the thought; he was too busily engaged in running at suicidal speed across the hot sand toward barren mountains where a ribbon of road showed through quivering air.

CHAPTER VIII
THE DARKNESS

Darkness; and red fires that seemed whirling about him as his body twisted in air. To Dean Rawson, plunging down into the volcano's maw, each second was an eternity, for, in each single instant, he was expecting crashing death.

Then he knew that long arms were wrapped about him, holding him, supporting him, checking his downward plunge ... and at last the glassy walls, where each bulbous irregularity shone red with reflected light, moved slowly past. And, after more eons of time, a rocky floor rose slowly to meet him.

His body crashed gently; he was sprawled face downward on stone that was smooth and cold. The restraining arms no longer touched him.

He lay motionless for some time, his mind as stunned and uncomprehending as if he had truly crashed to death upon that rocky floor. Then, at last, he forced his reluctant nerves and muscles to turn his body till he lay face upward.

Darkness wrapped him as if it were the soft swathing of some black cocoon. The world about him was at first a place of utter night-time blackness; and then, far above him, there shone a single star ... until that feeble candle-gleam, too, was snuffed out.

A hand was gripping his shoulder; it seemed urging him to arise. He felt each separate finger—long, slender, like bands of steel. The nail at each finger-end was more nearly a claw, the whole hand a thin, clutching thing like the foot of some giant ape. And, even as he shrank involuntarily from that touch, Rawson wondered how the creature could reach out and grip him so surely in the dark. But he came to his feet in response to that urging hand.

The night was suddenly sibilant with eery, whistling voices. They came from all sides at once; they threw themselves back and forth in endless echoes. To Rawson it was only a confused medley of conflicting sounds in which no one voice was clear. But the creature that held him must have understood, for he heard him reply in a sharp, piercing tone, half whistle, half shriek.

What had happened? Where was he? What was this thing that pushed him, stumbling, along through the dark? With all his tumultuous questioning he knew only one thing definitely: that it would be of no use to struggle. He was as helpless as any trapped animal.

He was inside the earth, of course; he had fallen he had no least idea how far; and, in some strange manner, this long-armed thing had supported him and eased him gently down. But what it meant or what lay ahead were matters too obscure for him to try to see clearly.

He held his hands protectingly before him while the talons gripping into his shoulder hurried him along. He stumbled awkwardly as his foot struck an obstruction. He would have fallen but for the grip that held him erect.

For that creature, whatever it was, the darkness held no uncertainty. He moved swiftly. His shrill shriek and the jerk of his arm both gave evidence of his astonishment that his captive should walk so blunderingly.

Then it seemed that he must have comprehended Rawson's blindness. A green line of light passed close behind Dean's head. It was cold—there was no radiant warmth—but, when it struck the face of a wall of stone some twenty feet away, the solid rock turned instantly to a mass of glowing yellow-red.

The cold green ray swung back and forth, leaving a path of radiant rock behind it wherever it touched. And the rock was hot! Once the green light held more than an instant in one place, and the rock softened at its touch, then splashed and trickled down to make a fiery pool.

Abruptly Rawson was able to see his surroundings. Also, he knew the source of the red glow that had seemed like volcanic fires. There had been others like his captor; they had been down below, and had played their flames upon the rocks deep in the volcano. It was thus that they made light.

With equal suddenness, and with terrible clearness, Dean found the answer to one of his questions. He wrenched himself about to stare behind him at the creature that held him in its grip. And, for the first time, the wild experience became something more than an unbelievable nightmare; in that one horrifying instant he knew it was true.

Only a few minutes before, he had been walking across the cindery sand of the crater top, walking under the stars and the dark desert sky— Dean Rawson, mining engineer, in a sane, believable world. And now...!

He squinted his eyes in the dim light to see more plainly the beastly figure, more horrible for being so nearly human. He had seen them briefly up above; the closer view of this one specimen of a strange race was no more pleasing. For now he saw clearly the cruelty in the face. It was there unmistakably, even though the face itself, under less threatening circumstances, might have been a ludicrous caricature of a man's.

Red and nearly naked, the creature stood upright, straps of metal about its body. It was about Rawson's height; its round, staring eyes were about level with his own, and each eye was centered in a circular disk of whitish skin. The light went dim for a moment, and Dean, staring in his turn, saw those other huge eyes enlarge, the white covering of each drawing back like an expanding iris.

Some vague understanding came to him of the beast's ability to see in the dark. They used these red-hot stones for illumination, but this thing had seemed to see clearly even when the stones had ceased to glow. And again, though indistinctly, Dean knew that those eyes might be sensitive to infra-red radiations — they might see plainly by the dark light that continued to flood these rocky chambers, though, to him, the rocks had gone lightless and black.

Even as the quick thoughts flashed through his mind, he was thinking other thoughts, recording other observations.

The rest of the face was red like the body; the head was sharply pointed, and crowned with a mass of thin, clinging locks of hair. The mouth, a round, lipless orifice, contracted or dilated at will; from it came whistling words.

Out of the darkness, giant things were leaping. They clutched at Rawson, while the first captor released his hold and drew back. Taller, these newcomers were, bigger, and different.

In the red light from the hot rocks Dean saw their faces, in which were owl eyes like those of the first one, but yellow, expressionless and stupid. Their great bodies were yellow: their outstretched hands were webbed.

For one instant, as Rawson's hand touched his pistol in its holster, a surge of fighting rage swept through him. His whole being was in a spasm of revolt against all this series of happenings that had trapped him; he wanted to lash out regardless of consequences. Then cooler judgment came to his aid.

Other figures, with faces red and ugly, expressive of nameless evil, were gathered beside the one who still played the jet of cold fire upon the walls. Like him they were naked save for a cloth at the waist and the metal straps encircling their bodies. They, too, had flame-throwers — he saw the long metal jets and their lava tips. Yet the temptation to fire into that group as fast as he could pull trigger was strong upon him.

Instead he allowed these other giant things to grip him with their webbed hands and lead him away.

The wavering light had shown many passages through the rock. Glazed, all of them. Either they had been blown through molten rock which had then solidified to give the glassy surfaces, or else — and this seemed more likely — the flame-throwers had done it. Rawson, scanning the labyrinth for some recognizable strata, had a quick vision of these caverns being cut out and enlarged, and of their walls melted just as they were being melted now — melted and hardened again innumerable times by succeeding generations of red and yellow-skinned men.

Yes, they were men. He admitted this while he walked unresistingly between two of the giants. Another went before them and lighted the way with the green ray of a flame-thrower on the melting rock. These were men — men of a different sort. Evolution, working strange changes underground, had made them half beasts, diggers in the dark, mole-men!

They were passing through a long tunnel that went steadily down. Cross passages loomed blackly; ahead of them the leader was throwing his flame upon the walls of a great vault.

Rawson had ceased to take note of their movements. What use to remember? He could never escape, never retrace his steps.

He tried to whip up a faint flicker of hope at thought of Smithy. Smithy had seen him go, had seen the red mole-men, of course. And he had got away — he must have got away! He would go for help....

But, at that, he groaned inwardly. Smithy would go for help, and then what? He would be laughed out of any sheriff's office; he would be locked up as insane if he persisted. Why should he persist — for that matter, why should he go at all? Smithy would not believe for a single minute that Rawson was still alive.

His thoughts ended. Webbed hands, wrapped tightly about his arms, were thrusting him forward into a great room. The green flame had been snapped off. One last hot circle on the high wall showed only a dull red. But before it faded, Dean saw dimly the outlines of a tremendous cavern. He saw also that these walls were unglazed, raw; they had never been melted.

Below the rough and shattered sides heaps of fragments were piled about the room.

Fleetingly he saw the shadowed details; then darkness swallowed even that little he had seen. Clanging metal told of a closing door; a line of red outlined it for an instant to show where it was welded fast. He was a prisoner in a cell whose walls were the living rock.

For a long time he stood motionless, while the heavy darkness pressed heavily in upon his swimming senses; he sank slowly to the floor at last. He was numbed, and his mind was as blank as the black nothingness that spread before his staring eyes. In a condition almost of coma, he had no measure or count of the hours that passed.

Then a fever of impatience possessed him; his thoughts, springing suddenly to life, were too wildly improbable for any sane mind, were driving him mad. He forced himself to move cautiously.

On the floor he had seen burnished gold, shining dully as he entered. There had been a thick vein of yellow in the rock. The floor, at that place, was rough beneath his feet, as if the hot metal had been spilled.

His hands groped before him as he remembered the heaps of rock fragments. Then his feet found one of them stumblingly, and he turned and moved to one side. He remembered having seen a dim shape off there that had made a straight slanting line. His searching hands encountered the object and kept him from walking into it.

The feeling of helplessness that drove him was only being increased by his blind and blundering movements. He told himself that he must wait.

Silently he stood where he had come to a stop, hands resting on the object that barred his way — until suddenly, stiflingly, his breath caught in his throat. Some emotion, almost too great to be borne, was suffocating him.

Slowly he moved his hands. Inch by inch he felt his way around the smooth cylinder, so hard, so coldly metallic. Then, with a rush, he let his hands follow up the slanting thing, up to a rounded top, to a heavy ring and a shackle that was on the end of a cable, thin and taut. And, while his hands explored it feverishly, the metal moved!

He clung to the smooth roundness as it slipped through his hands. It was the bailer, part of his own equipment. That slender cable reached up, straight up to the world he knew. And Smithy was there — Smithy was hoisting it!

He clung to the cylinder desperately. The bore, at this depth, had been reduced to eight inches; the bailer fitted it loosely. And Rawson cursed frantically the narrow space that would let this inanimate object return but would hold him back, while he wrapped his arms about the cold surface of the metal messenger from another world.

It lifted clear, then settled back. This time it dropped noisily to the floor. And suddenly Dean was tearing at the ring on one of the swollen fingers of his left hand.

It came free at last; it was in his hand as the cable tightened again. Swiftly, surely, he worked in the darkness to jam the ring through the shackle at the bailer's top. Then the bailer lifted, clanged loudly as it entered the shattered bore in the rocks above, and scraped noisily at the sides. The sound rose to a rasping shriek that went fainter and still fainter till it dwindled into silence.

But Dean Rawson, standing motionless in the darkness of that buried vault, dared once more to let himself think and *feel* as he stared blindly upward.

Up there Smithy was waiting. Smithy would know. And with Smithy fighting from the outside and he, Rawson, putting up a scrap below.... He smiled almost happily as his hand rested upon his gun.

Hopeless? Of course it was hopeless. No use of really kidding himself — he didn't have the chance of a pink-eyed rabbit.

But he was still smiling toward that dark roof overhead as the outlines of a metal door grew cherry red. They were coming for him! He was ready to meet whatever lay ahead....

CHAPTER IX
A SUBTERRANEAN WORLD

The metal plate that had sealed him in this tomb fell open with a crash. Beyond it the passageway was alive with crowding red figures. Above their heads the nozzles of a score of flame-throwers spat jets of green fire. Rawson drew back in sudden uncontrollable horror as they came crowding into the room.

The familiar feel of the bailer's cold metal had given him a momentary sense of oneness with his own world. Now this inrush of hideous, demoniac figures beneath the flare of green flames was like a fevered vision of the infernal regions come suddenly to actuality.

Rawson retreated to the shattered, rocky wall and prepared for one last fight, until he realized that the evil black eyes in their ghastly circles of white skin were fixed upon him more in curiosity than in active hatred.

They formed a semicircle about him—a wall of red bodies, whose pointed heads were craned forward, while an excited chatter in their broken, whistling speech filled the room with shrill clamor. Then one of them pointed above toward the open shaft that Rawson had drilled, the shaft up which the bailer had gone. And again their voices rose in weird discord, while their long arms waved, and red, lean-fingered hands pointed.

Only a moment of this, then one of them gave an order. Two of the red figures came toward Rawson where he was waiting. They were unarmed. They motioned that he was to go with them. And Dean, with a helpless shrug of his shoulders, allowed them, one on each side, to take him by the arms and hurry him through the open door. Two others went ahead, the green jets of flame from their weapons lighting the passage.

The system of communicating tunnels seemed at first only the vents and blow-holes from some previous volcanic activity. And yet, at times they gave place to more regular arrangement that plainly was artificial. The air in them was pure, though odorous with a pungent tang which Dean could not identify. Through some of the passages it blew gently with uncomfortable warmth.

The guard of wild red figures hurried him along through a vast world of caverns and winding passages which seemed one great mine. The richness of it was amazing. Dean Rawson was a man, a human being, facing death in some form which he could not yet know, and, so fast had his wild experiences crowded in upon him, he seemed numbed to all normal emotions; yet through it all the mind of the engineer was at work, and Dean's eyes were flashing from side to side, trying to see and understand the ever-changing panorama of a subterranean world.

Mole-men, both red and yellow, were everywhere. But it was apparent at a glance that the yellow giants were a race of toilers—slaves, driven by the reds.

Their great bodies glowed orange-colored with the reflected heat of the blasts of flame used to melt the metals from their ores. Gold and silver, other metals that Rawson could not distinguish in the half light—the glow of the molten stuff came from every distant cave that the passages opened up.

The sheer marvel of it overwhelmed him. His own danger, even the death that waited for him, were forgotten.

A world within a world—and who knew how far it extended? Mole-men, by scores and hundreds, the denizens of a great subterranean world, of which his own world had been in ignorance. Here was civilization of a sort, and now the barriers that had separated this world from the world above had been broken down; the two were united. Suddenly there came to Rawson's mind a flashing comprehension of a menace wild and terrible that had come with the breaking of those barriers.

They were passing through a wider hall when the whistling chatter of Dean's escort ceased. They were looking to one side where a cloud of smoke had rolled from a slope beyond. One of the red figures staggered, choking, from the cloud. Two yellow mole-men followed closely after.

The red mole-man was unarmed; each yellow one had a flame-thrower that was now so familiar a sight to Dean. His own escort was silent; they had halted, watching those others expectantly.

In the silence of that rocky room the single red one whistled an order. One of the two yellow men placed his weapon on the floor. Another shrill order followed, and the remaining worker, without a moment's hesitation, turned the green blast of his own projector upon his comrade.

It was done in a second—a second in which the giant's shriek ended in a flash of flame for which his own flesh was the fuel. A wisp of drifting smoke, and that was all. And the red creatures who had Rawson in their charge, after a moment of silence, filled the room with shrill-voiced pandemonium, while they shrieked their approval of the spectacle.

But Dean Rawson's lips were forming half-whispered words, so intently was he thinking the thoughts. "The damned red beast! That poor devil's flame hit some sulphur, I suppose—burned it to SO_2—then he got his!"

But, even while he searched his mind for words to describe the evil of this red race, he was realizing another fact. These yellow giants, countless thousands of them, perhaps, were held in subjection by their red masters. They would do as they were told. Dimly, vaguely, through his horrified mind, came the picture of a horde of red and yellow beasts turned loose upon the world above.

There were fears now which filled Dean Rawson, shook him with horrors as yet only half comprehended. But the fears were not for himself, one solitary man in the grip of these red beasts—he was fearing for all mankind.

His guard was hurrying him on, but now Dean hardly saw the scenes of feverish activity through which they passed. Another thought had come to him.

That shaft, the hole which he himself had drilled—what damage had it done? It was he who had broken down the barriers. His drill had told these beasts that there was other life above. It had guided them. They had realized that they were near to some other place where men worked and drove tunnels through the rocks. They had followed up these forgotten passages that led to the old craters, had ascended inside the volcano, made their way through the top and emerged into another world—a clean and sunlit world.

Now Rawson's eyes found with new understanding the activity about him.

The mining operations had been left behind. Here were branching passages, great cavelike rooms—a world within a world, in all truth. Throughout it, demoniac figures were hurrying, driving thousands of giant yellow slaves where the light shone sparkling from innumerable heaps of metal weapons—flame-throwers and others, the nature of which Rawson could not determine. And everywhere was the shouting and hurry as of a nation in the throes of war.

His speculations ended abruptly. They were approaching a room, a vast open place. High on the farther wall was a recess in the rock in which tongues of flame licked hungrily upward. The heat of the fires struck down in a ceaseless hot blast. Close to the fires, unmindful of the heat, a barbaric figure assumed grotesque and horrible postures, while its voice rose in echoing shrillness.

Below were crowding red ones who prostrated themselves on the rocky floor.

"Fire worshipers!" The explanatory thought flashed through Dean Rawson's mind. "Here was one of their holy places, a place of sacrifice, perhaps, and he was being taken there, helpless, a captive!"

CHAPTER X
PLUMB LOCO

The sheriff of Cocos County was reacting exactly as Rawson had anticipated. Smithy stood before him, a disheveled Smithy, grimy of face and hands. He had made his way to the highway and caught a ride to the nearest town, and now that he had found Jack Downer, sheriff, that gentleman leaned back in his old chair behind the battered desk and regarded the younger man with amused tolerance.

"Now, that's right interesting, what you say," he admitted. "Tonah Basin, and the old crater, and red devils settin' fire to everything. I've heard some wild ones since this Prohibition went into effect and some of the boys started makin' their own, but yours sure beats 'em all. Guess likely I'll have to take a run up Tonah way and see what kind of cactus liquor they're makin'."

"Meaning I'm drunk or a liar." Smithy's voice was hot with sudden anger, but the sheriff regarded him imperturbably.

"Well, I'd let you off on one count, son. You do look sort of sober."

Smithy disregarded the plain implication and fought down the anger that possessed him.

"May I use your phone, Mr. Downer?" he asked.

He called the office of Erickson and his associates in Los Angeles and told, as well as he could for the constant interruptions from his listener, the story of what had occurred. And Mr. Erickson at the other end of the line, although he used different words, gave somewhat the same reply as had the sheriff.

"I refuse to listen to any more such wild talk," he said. "If our property has been destroyed, as you say, there will be an accounting, you may be sure of that. And now, Mr. Smith, get this straight, you tell Rawson, wherever he is hiding, to come and see me at once."

"But I tell you he has been captured," said Smithy desperately. "He's gone."

"I rather think we will find him," was the reply. "He had better come of his own accord. His connection with us will be severed and all drilling operations in Tonah Basin will be discontinued, but Mr. Rawson will find that his responsibility is not so easily evaded."

The sheriff could not have failed to realize the unsatisfactory nature of the conversation; he must have wondered at the satisfied grin that spread across Smithy's tired face.

"Do you mean you're through?" he demanded. "You're abandoning Rawson's work?"

"Exactly," was Mr. Erickson's crisp response.

Smithy, as the telephone clicked in his ear, turned again to the sheriff. "That unties my hands," he said cryptically. "One more call, if you please."

Then to the operator: "Get me the offices of the Mountain Power and Lighting Corporation in San Francisco. I will talk with the president."

The sheriff of Cocos County chuckled audibly. "You'll talk to the president's sixteenth assistant secretary, son," he told Smithy. "And I take back what I said before—now I know you're plumb loco. By the way, son, it costs money for telephone calls like that. I hope you ain't, by any chance, overlookin'—"

But Smithy was speaking into the telephone unmindful of the sheriff's remarks.

"Is Mr. Smith in his office?" he was inquiring. "Yes, President Smith.... Would you connect me with him at once, please? This is Gordon Smith talking."

"Hello, Dad," he said a moment later. "Yes, that's right. It's the prodigal himself. Now, listen, Dad, here's something important. Can you meet me in Sacramento and arrange for us to see the Governor—get his private, confidential ear? I'll beat it for Los Angeles—charter the fastest plane they've got...."

There was more to the conversation, much more, although Smithy refrained from giving details over the phone. An operator was breaking in on the conversation as he was about to hang up.

"Emergency call," the young woman's voice was saying. "We must have the line at once."

Smithy handed the telephone to the sheriff. "Someone's anxious to talk to you," he said. He searched his pockets hurriedly, found a ten-dollar bill which he laid on the sheriff's desk. "That will cover it," he said with a new note in his voice. "Perhaps you're not just the man for this job, sheriff. It's going to be a whole lot too hot for you to handle."

He had turned quickly toward the door, but something in the sheriff's excited voice checked him. "Burned? Wiped out, you say?"

Halfway across the room Smithy could hear another hoarse voice in the telephone. The sheriff repeated the words. "Red devils! They wasn't Injuns? The whole town of Seven Palms destroyed!"

"I thought," said Smithy softly to himself, "that we'd have to go down *there* to find *them*, and instead they're out looking for us. Yes, I think this will be decidedly too hot for you to handle, sheriff." He turned and bolted out the door.

An attentive audience was awaiting Gordon Smith on his arrival in Sacramento. Smithy's father was not one to be kept waiting even by the Governor of the state. Also, Smithy was coming from the Tonah Basin region, and the news of the destruction of the desert town of Seven Palms had preceded him. Even the swift planes of the Coastal Service could not match the speed of the radio news.

There were only two men in the room when Smithy entered. One of them, tall, heavily built, as square-shouldered as Smithy, came forward and put his two hands on the young man's shoulders. Their greetings were brief.

"Well, son?" asked the older man, and packed a world of questioning into the interrogation.

"O. K., Dad," said Smithy simply.

His father nodded silently and turned to the other man. "Governor, my son, Gordon. He got tired of being known as the 'Old Man's son'—started out on his own—not looking for adventure exactly, but I judge he has found it. He's got something to tell us."

And again Smithy told his wild, unbelievable tale. But it was not so incredible now, for, even while Smithy was talking, the Governor was glancing at the report on his desk which told of the destruction of the little town of Seven Palms.

"I can't tell you what it means," Smithy concluded. He paused before venturing a prediction which was to prove remarkably accurate. "But I saw them—I saw them come up out of the earth, and I'm betting there are plenty more where they came from. And now that they've found their way out, we've got a scrap on our hands. And don't think they're not fighters, either. They're armed—those flame-throwers are nothing we can laugh off, and what else they've got, we don't know."

He leaned forward earnestly across the Governor's desk. "But that's your job," he said. "Mine is to find Dean Rawson. He's alive, or he was. He sent up his ring as proof of it. I've got to find him—I've got to go down in that pit and I want your help."

CHAPTER XI
THE WHITE-HOT PIT

How far his guard of wild, red man-things had taken him Dean Rawson could not know. Many miles, it must have been. And he knew that the air had grown steadily more stiflingly hot. But the heat of those long tunneled passages was like a cool breeze compared with the blasting breath of the room into which he was plunged.

It seared his eyeballs; it struck down from the tongues of flame that played in red fury in the recess high up on the farther wall. And the vast room, the fires, the hundreds of kneeling figures, all blurred and swam dizzily before him.

The hot air that he breathed seemed crisping his lungs. Vaguely, for the stupefying, brain-numbing heat, he wondered at the figure he saw dimly in its grotesque posturing close to the flames. And the hundreds of others—how could they live? How could he himself go on living in this inferno?

They had been chanting in unison, the kneeling red ones. Dean heard the regular beat of their repeated words change to an uproar of shrill, whistling voices. But he could neither see nor hear plainly for the unbearable, suffocating heat.

The clamor was deafening, confusing; it echoed tremendously in the rocky room and mingled with the steady, continuous roar of the flames. The mass of bodies that surged about him made only a blurring impression; he tried to make himself see clearly. He must fight—fight to the last! Only this thought persisted. He was striking out blindly when he knew that his red guard had cleared a way through the mob and was dragging him forward.

He knew when they reached the farther wall. Somewhere above him was the deep-cut niche in which the fires roared. And then, when again he could see from his tortured eyes, he found directly ahead another doorway in the solid rock. Beyond it all was black; it gave promise of coolness, of relief from the stifling air of the room. Red hands were thrusting him through.

The burst of water, icy cold, that descended upon him from above shocked him from the stupor that claimed his senses. He was drenched in an instant, strangling and gasping for breath. But he could think! And, as the lean hands seized him again and hurried him forward, he almost dared to hope.

To his eyes the passageway was a place of utter darkness, but the red ones, their great owl eyes opened wide, hurried him on. His stumbling feet encountered a flight of steps. With the red guard he climbed a winding stair where the tunnel twisted upward.

That icy deluge had set every nerve aquiver with new life. He hardly dared ask himself what might lie ahead. Yet he had been saved from that mob; it might be his life would be spared, that in some way he could learn to communicate with these people, learn more of this subterranean world—which must be of tremendous extent. Without any sure knowledge of their plans, he still was certain in his own mind that they intended to swarm out upon the upper world. He might even be able to show them the folly of that.

A thousand thoughts were flashing through his mind when the tunnel ended. Beyond a square-cut opening the air was aglow with red. An ominous thunder was in his ears. Then a score of hands lifted him bodily and threw him out upon a rocky floor that burned his hands as he fell.

Heat, blistering, unbearable, beat upon him. He was wrapped in quick-rising clouds of steam from his wet clothes.

The platform ended. Far below was a sea of red faces, grotesque and horrible, where each held two ghastly white disks, and at the center of each disk a mere pinpoint eye.

He saw it all in the instant of his falling—the inhuman, shrieking mob, the blast of hot flame not forty feet away at the back of the rocky niche, and, between himself and the flame, a giant figure that leaped exultantly, while its body, that appeared carved from metallic copper, reflected the red fires until it seemed itself aflame.

Dean knew in the fraction of a second while he scrambled to his feet, that the great room had gone silent. The roaring of the flames ceased; even the clamor of shrill voices was stilled. He had thrown one arm across his face to shield his eyes; the heat still poured upon him like liquid fire. But his instant decision to throw himself out and down into the waiting mob was checked by the sudden stillness.

To open his eyes wide meant impossible torture, yet he forced himself to peer through slitted lids beneath the shelter of his arm.

The flame was gone. Where it had been was a wall of shimmering red rock above a gaping throat in the floor, whose rim was quivering white with heat. Here the blast from some volcanic depth had come.

Then he saw it, saw the great coppery figure leaping upon him—and saw more plainly than all this the end that had been prepared for him.

Fire worshipers! Demons of an under world paying tribute to their god. And he, Dean Rawson, was to be a living sacrifice, cast headlong to that waiting, white-hot throat!

The coppery giant was upon him in the instant of his realization. Somehow in that moment Dean Rawson's wracked body passed beyond all pain. With the inhuman, maniacal strength of a man driven beyond all reason and restraint he tore himself half free from those encircling arms and drove blow after blow into the hideous face above him.

Only his left arm was free. That, too, was clamped tightly against his body an instant later.

The giant had been between him and the glowing rocks. Now he felt himself whirled in air, and again the blast of heat struck upon him. He was being rushed backward; and there flashed through his mind, as plainly as if he could actually see it, the scintillant whiteness of that hungry throat.

He tried to lock his legs about the big body to prevent that final heave and throw that would end a ghastly ceremony. The rocks were close, their radiant heat wrapped about him like a living flame. Abruptly his strength was gone—the fight was over—he had lost! His heart sent the blood pounding and thundering to his brain; his lungs seemed on fire.

The high priest of the red ones had his priestly duty to perform—the sacrifice must be offered. But even the high priest, it would seem, must have been not above personal resentment. Sacrilege had been done—a fist had smashed again and again into the holy one's face. This it must have been that made him pause, that brought one big hand up in a grip of animal rage about Dean's throat.

Only a moment—a matter of seconds—while he vented his fury upon this white-skinned man who had dared to oppose him. Dean felt the hand close about his throat. So limp he was, so drained of strength, he made no effort to tear it loose. He was *dead*—what mattered a few seconds more or less of life? And then a thrill shot through him as he knew his right hand was free.

That hand made fumbling work of drawing a gun from its smoking, leather holster. He could hardly control the numbed, blistered fingers, yet somehow he crooked one about the trigger; and dimly, as from some great distance, he heard the roar of the forty-five.... Then, from some deep recess within him, he summoned one last ounce of strength that threw him clear of the falling body.

Instinctively he had heaved himself away from the fiery rocks; the same effort had sent his big coppery antagonist staggering, stumbling, backward. And Dean, sprawled on the stone floor, whose heat where he lay was just short of redness, heard one long, despairing shriek as the giant figure wavered, hung in air for a moment in black outline against the fierce red of a rocky wall above a white-hot pit, then toppled, pitched forward, and vanished.

Sick and giddy, he forced himself to draw his body up on hands and knees. Then he straightened, came to his feet, and staggered forward.

Below him was pandemonium. The sea of faces wavered and blurred before his eyes. From a distant archway other figures were coming. He saw the gleam of metal, heard the wild blare of trumpets, and knew that the hundreds of red ones below him were standing stiffly, both hands raised upright in salute as another barbaric figure entered. The air was clamorous with a shrill repeated call. "Phee-e-al!" the red ones shrieked. "Phee-e-al!"

But Rawson did not wait to see more. Behind him, the flames that had been fed with human flesh—if indeed these red ones were human—roared again into life. He had returned the pistol to its holster when first he came to his feet; his weak hands had seemed unable to hold it. And now his two hands were thrust outward before him as he staggered blindly toward the tunnel mouth.

It was where he had emerged upon the platform. His reaching hands found the side entrance where the stairs led down to the main hall. In the darkness he made his way past. Stumbling weakly he pushed on down the long tunnel whose floor slanted gently away.

Ahead of him was a light. The comparative coolness of these rocks had served to revive him somewhat. He had no hope of escape, yet the light seemed comforting, somehow.

He stopped. His stinging eyes were wide open. He stared incredulously at the glowing spot on a distant wall, where a flame must have touched, and at the figure beneath it.

The figure of a woman! A young woman, tall, slender, fair-haired, whose skin was white, a creamy white, whiter than snow.

A woman? It was a mere girl, slender and beautiful, her graceful young body poised as if, in quick flight, she had been caught and held for a moment of stillness.

What was she doing here? His exhausted brain could not comprehend what it meant. He had seen women of the Mole-men tribe mingling with the men. Like them their heads were pointed, their faces grotesque and hideous. Rawson gave an inarticulate cry of amazement and staggered forward.

Between him and the distant figure a crowd of Reds swarmed in. They came from a connecting passage. Above their heads the lava tips of flame-throwers were spitting jets of green fire. Every face was turned toward him at his cry.

Beyond them the white figure vanished. Dean, leaning weakly against the wall, told himself dully that it had been a phantom, a product of his own despairing brain and his own weakness. Then that weakness overcame him; and the red Mole-men, their white and hideous eyes, the threatening jets of green flame, all vanished in the quick darkness that swept over him....

CHAPTER XII
DREAMS

The black curtain of unconsciousness which descended so quickly upon Rawson was not easily thrown off. For hours, days or weeks—he never knew how long he lay in the citadel of the Reds—it was to wrap him around.

Nor was his waking a matter of a moment. Many and varied were the impressions which came to him in times of semiconsciousness, and which of them were realities and which dreams, he could not tell.

He was being tortured with knives, lances tipped with pain that dragged him up from the black depths in which he lay. Dimly he realized that his clothes were being stripped from him and that the piercing knives were none the less real for being only the touch of hands and rough cloth upon his blistered body. Then from head to foot he was coated with a substance cool and moist. The pain died to a mere throbbing and again he felt himself sinking back into unconsciousness.

There were other visions, many others, some of them plain and distinct, some blurred and terrifying to his fevered brain trying vainly to bring order and reason into what was utterly chaotic.

Once a bedlam of shrieking voices roused him. He tried to open his eyes, whose lids were too heavy for his strength. And by that he knew he was dreaming. Yet from under those lowered lids he seemed to see a wild medley of red warriors, their faces blotched and ghastly in the green light of their weapons. They were carrying a charred body which they threw heavily upon the floor beside him as if to compare the two. He saw the face which the flames had not touched, the face of Jack Downer—Downer, the sheriff of Cocos County. His sandy hair had been scorched to the scalp.

Dreams ... and the steady beat of metal-shod feet of marching men. He saw them passing some distance away. The repeated *thud-thud* of metal on stone echoed maddeningly through his brain for hours.... Dreams, all of them.

And once there came to him a vision which beyond all doubt was unreal.

Silence had surrounded him. For what seemed hours not one of the red mole-men had come near. And then, in the silence, he heard whisperings and the sound of stealthy feet; and, for a moment, the same white figure that had met him in his flight stood where he could see.

Only the merest trace of dim light relieved the utter darkness of the room. The girl's figure was ghostly, unreal. Yet he saw the dull sparkle of jeweled breast-plates against her creamy white skin. Loose folds of cloth were gathered about her waist; her golden hair was drawn back except for vagrant curls that only accentuated the perfect oval of her face.

There were others with her, dim shapes of men; how many Rawson could not tell. They looked down at him, whispering softly, excitedly, amongst themselves; but their words were like nothing he had ever heard.

For an instant Dean felt his stupefied mind coming almost to wakefulness. Phantom figures, ghostly and unreal—but the faces were human, and the eyes looked down upon him pityingly. He tried to rouse himself, tried to call out, then settled limply back, for the girl was speaking—or he was catching her thoughts. It seemed almost that he heard her whispered words:

"They take him to *Gevarro*, to the Lake of Fire which never dies! Gor told me—he overheard their plans. But, by the Mountain I swear...." Then footsteps echoed in a far-off passage, and the white ones vanished like drifting smoke.

Dreams, all of them. Yet the time came when Dean knew that he was awake—knew too that further experiences awaited him in this demoniac land.

Again red guards came. The wicked breath of their weapons filled the great room where Rawson had been with green, flickering light. Dean, dragged to his feet, was unable to stand. One of the giant yellow workers came forward at a whistled order and held him erect. Another brought a bowl carved from rock crystal and filled with a liquid golden-green with reflected light. He put it to Rawson's lips and with the first touch Dean knew that he must have been filled with a burning thirst beyond anything he had ever known. He gulped greedily at the liquid, drained the bowl to the last drop, then marveled at the thrilling fire of strength that flowed through him.

"Wine," he thought, "wine of the gods — or devils." He came to himself with a start. He knew that he was naked and that his body was encased in a coating of stiff gray plaster. It was this that prevented his arms and legs from flexing.

Another order and the giant worker picked him up in his arms and carried him where the others led to a distant room. A stream trickled through a cut in the rocky floor. At the center of the room was a pool. Unable to resist, Dean felt the giant arms toss him out and down.

The water was warm. At its first touch the hard plaster melted like snow. Sputtering and choking for breath, Rawson came to the surface. He found he could move freely, then reaching hands hauled him out upon the floor, and through all his dread he found time to marvel at his own firm muscles and the healthy white of his skin that had been seared and blistered.

He obeyed when the red guards pointed and motioned him into a dark passageway. He tried to keep up with them as they hurried him on. Evidently his pace was too slow, for again the big worker picked him up, swung him into the air and seated him firmly on one broad shoulder, and, with red guards ahead and behind them, hurried on.

To find himself a child in the hands of this big yellow man was disconcerting. To be calmly lugged off was almost humiliating. No one who was not a good sport could have grinned as Rawson did at his own predicament.

"Not exactly a triumphal procession," he told himself, then his lips set grimly. "They've got my gun," he thought, "and now, whatever comes, all I can do is stand and take it. Still, they've saved my life. But what for?"

Always the way led downward, and Rawson, perched on his strange, half-human steed, let his gaze follow up every branching tunnel and widespread cave. Not all of these were as dark as the broad thoroughfare they followed. In some, strange lights glowed, and Rawson saw weird, towering plant growths that yellow workers were harvesting.

Life, life, everywhere, and seemingly this underground world was endless.

Troops of red warriors passed them, upward bound. The dancing flames of their weapons, where occasional ones were in action, glowed from afar. They bobbed and waved like green fireflies as the Mole-men came on at a half-run.

"And this means trouble up top," he thought. "There's going to be hell to pay up there."

But workers, fighters, everyone they met stood aside to let the red guard pass. Again Rawson heard the strange word or call that had come to him in the temple of fire. One of the guides would give a whistling call that ended in the same strange shrill cry of "Phee-e-al," and instantly the way was cleared.

A wild journey, incredible, unreal. Rawson, as he met the countless staring white eyes of the creatures they passed, found his thoughts wandering. He had had wild dreams. Surely this was only another in that succession of phantom pictures. Then, seeing the cold, implacable hatred in those staring eyes, he would be brought back with sickening abruptness to a full knowledge of his own hopeless situation.

"Gevarro, the lake of fire which never dies" — what was it the white ones had said? But no, that certainly was a dream like that other in which he had seemed to see the charred body of a man, the sheriff who had called to see him at his camp in Tonah Basin.

Dreams — reality — his brain was confused with the wild kaleidoscope of unbelievable pictures.

He was suddenly aware that through it all he had been mentally tabulating their route, remembering the outstanding features when there was light enough to see. He knew that unconsciously his mind had been thinking of escape. Wilder than all the other visions, he had been picturing himself retracing his route, alone, free. He did not know that he had laughed aloud, harshly, hopelessly, until he saw the curious eyes of his red guard upon him.

"Yes," he told himself in silent bitterness, "I could find my way back, if...."

The guard had swung off from the great tunnel which must have been one of the main thoroughfares of the Mole-men's world. They crowded through a narrower passage and again Rawson found himself in one of the great, high-ceilinged caves like the others he had seen. But unlike the others this was brightly lighted.

Massive limestone formation. His eyes squinted against the glare and caught the character of the rock before he was able to distinguish details, and in the black limestone big disks of gray mineral had been set. Jets of flame played upon them and turned them to blazing, brilliant white.

The big yellow Mole-man who had carried him dropped him roughly to the floor and backed away. About him the red guard was grouped. Rawson caught a glimpse of hundreds of other thronging figures. The crowd about him separated. A space was cleared between him and the farther end of the room, a lane lined on either side by solid masses of savage Reds. And beyond them, more barbaric than any figure in the foreground, was another group.

Across the full width of the room a low wall was raised three or four feet from the floor. It was capped with rude carvings. The whole mass gleamed dully golden in the bright light. Beyond the wall in semicircular formation, resembling a grouping of bronze statues, were men like the one with whom Rawson had fought. Priests, tenders of the fires. He knew in an instant that here were more of the red one's holy men. They stood erect, unmoving. At their center was another seated man-shape that might have been cast from solid gold.

His naked body was yellow and glittering, contrasting strongly with the black metal straps like those the warriors wore. On his head a round, sharply-pointed cap was ablaze with precious stones.

Rawson took it all in in one quick glance. He knew that those copper bodies were not encased in metal, for the flesh of the one he had fought with had sunk under his blows. Their skin was coated with a preparation, heat resistant without a doubt, and the golden one must have been treated in somewhat the same way.

His thoughts flashed quickly over this. It was the face of that seated figure that riveted his attention, a white face, milk-white, so white it seemed almost chalky!

For one breathless second Rawson was filled with a wordless hope. Those white ones of his dream had looked upon him with kindly eyes. They were human — men of another race, but men. Then beneath the chalky whiteness of the face he found the hideous features of the red Mole-men, and knew that the white color of the face was as false as that of the golden body.

But he was their leader. He was someone of importance. Rawson had started forward impetuously when he saw the figure rise. At the first motion the hands of every red one in the room were flung in air. They stood stiffly at salute. Even the priests' coppery arms flashed upward. And "Phee-e-al!" a thousand shrill voices were shouting. "Phee-e-al! Phee-e-al!"

Rawson stopped, then walked slowly forward, one defenseless, naked man of the upper world, between two living walls formed by men of a hidden race.

"Phee-e-al," he was thinking. "He's the one I saw coming into their temple back there. They got out of our way when they knew we were coming to see him. He's the big boss here, all right."

He did not pause in his steady, forward progress until his hands were resting upon the golden barrier. Strange thoughts were racing through his mind. Phee-e-al, he was facing Phee-e-al, king of a kingdom ten miles or more beneath the surface of the earth, a place of devils more real and terrible than any that mythology had dared depict. And he, Dean Rawson, a man, just one of the millions like him up there in a sane, civilized world, was down here, standing at a barrier of gold before a tribunal that knew nothing of justice or mercy.

Thoughts of communicating with them had mingled with other half-formed plans in his racing mind. Sign language—he had talked with the Indians; he might be able to get some ideas across. He met the other's fierce scrutiny fearlessly, then, waiting for him to make the first advance, let his gaze dart about at closer range. He could not restrain a start of surprise at sight of his own clothing, his pocket radio receiver and his pistol spread out on a metal stand.

They had been curious about them. Rawson took that as a good sign. Perhaps he had been mistaken in his interpretation of what he had seen. For himself, he could have no real hope, but it might be that the outpouring of these demons into his own world was a threat that lay only in his own imagination.

His eyes came back to meet that gaze which had never left him. The eyes were mere dots of jet in a white and repulsive face. The rounded mouth opened to emit a shrill whistled order.

In the utter silence of the great room one of the copper-skinned priests moved swiftly toward the rear. There were chests there, massive metal things afire with the brilliance of inlaid jewels. The priest flung one of them open with a resounding clang.

The room had been warm, and the chill which abruptly froze Rawson's muscles to hard rigidity came from within himself. Dreams! He had thought them dreams, those marching thousands, and the others who returned. He had dared to hope he might avert an invasion by this inhuman horde.

And now he knew his worst imaginings were far short of the truth. He saw clearly his own fate. For the priest returning was holding an object aloft, a horrible thing, a naked body, scorched and charred. And above it a head lopped awkwardly. The hair was sandy; half of it had been burned to the scalp in a withering flame. Below, staring from sightless eyes, was the face of the man who had once been sheriff of Cocos County.

CHAPTER XIII
"N-73 CLEAR!"

"You fly, of course?" demanded Governor Drake.

Smithy nodded. "Unlimited license—all levels."

They had spent the night in the executive mansion, and now the Governor had burst precipitately into the room where Smithy and his father had just finished dressing. The two had been deep in an earnest conversation which the Governor's entrance had interrupted.

"I am drafting you for service," said the Governor. "I want you to go out to Field Number Three. A fast scout plane—National Guard equipment—will be ready for you—"

He broke off and stared doubtfully at a paper in his hand, a radiophone message, Smithy judged. "I'm in a devil of a fix," the Governor exclaimed, after a pause. Then:

"I don't doubt your sincerity," he told Smithy. "Never saw you till yesterday, but your father's 'O.K.' goes a hundred per cent with me. Old 'J. G.' and I have been through a lot of scraps together." His frowning eyes relaxed for a moment to exchange twinkling glances with the older man.

"No, it isn't that," he added, "but...." Again he stared at the flimsy piece of paper.

"What's on your mind, Bill?" asked Smith senior. "That stuff the boy told us was pretty wild"—he laid one hand affectionately upon Smithy's shoulder—"but he's a poor liar, Gordon is, and, knowing his weakness, he usually sticks to the truth. And there's no record of insanity in the family, you know. If there's something sticking in your crop, Bill, cough it up."

And the Honorable William B. Drake obeyed. "Listen to this," he commanded, and read from the paper in his hand:

> "'Replying to your inquiry about the doings at Seven
> Palms. Some Indians did that job. No help needed. I can
> handle this. Posse organized and we are leaving right
> now.—Signed, Jack Downer, Sheriff, Cocos County.'"

"That sounds authentic," said Smithy drily. "I've met the sheriff."

"Now, if it *was* Indians that got tanked up and came down off the reservation, burned Seven Palms and cleaned up your camp—" began Governor Drake.

"It wasn't!" Smithy interrupted hotly. "I told you—" He felt his father's hand gripping firmly at his shoulder.

"Steady," said Smith, senior. "Let him talk, son."

"There's an election three months from now, J. G.," said the Governor, "and you know they're riding me hard. Let me make one false move—just one—anything that the opposition can use for a campaign of ridicule, and my goose is cooked to a turn."

Gordon Smith shook off his father's restraining hand and took one quick forward step. His face, even through the tan of the desert sun, was unnaturally pale.

"Election be dammed!" he exploded. "Dean Rawson has been captured by those red devils—he's down there, the whitest white man I ever met! I've been to the sheriff; now I've come to you! Do you mean to tell me there isn't any power in this state to back me up when—"

He stopped. There was a tremble in his voice he could not control.

"Good boy," said Governor Drake softly. "Now I know it's the truth. Yes, you'll be backed up, plenty, but for the present it will be strictly unofficial. Now pull in your horns and listen.

"You know the lay of the land. I want your help. Go out to Field Three; there'll be a man there waiting for you. Don't call him 'Colonel'—he's also strictly unofficial to-day. The sheriff and his posse will be there at Seven Palms inside an hour; I want you to be there, too, about five thousand feet up.

"Tell Colonel Culver—I mean Mr. Culver—your story; tell him everything you know. He'll be in charge of operations if we have to send in troops; he'll give you that private and unofficial backing I spoke of if we don't.

"Now get down there; keep your eye on the sheriff's crowd and see everything that happens!"

But Smithy's parting remark was to his father; it was a continuation of the subject they had been discussing before.

"You can buy at your own price," he said. "They've got rights to the whole basin. But they've quit; I'm not treating them to a double-cross."

And he added as he went out of the room: "Buy it for me if you don't want it yourself."

It was a two-place, open-cockpit plane that Smithy found had been set aside for him. Dual control—the stick in the forward cockpit carried the firing grip that controlled the slim blue machine guns firing through the propeller. Behind the rear cockpit a strange, unwieldy, double-ended weapon was recessed and streamlined into the fuselage. The scout seemed quite able to protect itself in an emergency.

Beside the plane a tall, slender man in civilian attire was waiting. He stuck out his hand, while the gray eyes in his lean, tanned face scanned Smithy swiftly.

"I'm Culver. Understand I'm to be your passenger to-day. How about it—can you fly the ship? Seven hundred and fifty DeGrosse motor—retractable landing gear, of course. She hits four-fifty at top speed—snappy—quick on the trigger."

Smithy shook his head dubiously. "Four-fifty—I'm not accustomed to that. But you can take the stick, Mr. Culver, if I get in a hurry and jump out and run on ahead. You see I'm used to my own ship, an *Assegai*—special job—does five hundred when I'm pressed for time."

The lean face of Mr. Culver creased into a smile. "You qualify," he said. "But keep your hands off the dead mule."

At an inquiring glance he pointed to the heavy, half-hidden weapon that Smithy had noticed. "Can't kick," he explained, "—hence 'dead mule.' It's the new Rickert recoilless; throws little shells the size of your thumb—but they raise hell when they hit."

"Sounds interesting." Smithy climbed into the rear cockpit and strapped himself in. "Show me how it works, then I won't do it."

A pistol grip moved under Culver's reaching hand and the strange weapon sprang from concealment like something alive. The pistol grip moved sideways, and the gun swung out and down, its muzzle almost touching the ground. Smithy was suddenly aware that a crystal above his instrument board was reflecting that same bit of sun-baked earth. A dot of black hung stationary at the crystal's center.

"That's your target." Culver's voice held all the pride of a child with a new toy, but he released the grip, and the ungainly gun swung smoothly back to its hiding place.

He settled himself in the forward cockpit. "You will find a helmet there," he said. "It's phone-equipped; you can tell me all about that wild nightmare of yours while we jog along."

The white beam from the despatcher's tower had been on them while they talked. Other planes were waiting on the field. Smithy smiled as he settled the helmet over his head. "For a strictly unofficial flight," he thought, "we're getting darned good service."

He taxied past a hangar where uniformed men pointedly paid them no attention. He swung the ship to the line as Airboard regulations required.

"N-73" was painted on the monoplane's low wings that seemed scraping the ground. "N-73 Clear!" the despatcher's voice radioed into Smithy's ears. Then the seven-hundred-and-fifty-horsepower DeGrosse let loose its voice as Smithy gunned her down the field.

Whatever doubts Colonel Culver may have had of Smithy's ability were dissipated as they made their way cautiously through the free-flying area under five thousand. Everywhere were mail planes, express and passenger ships taking off for the transcontinental day run, and private planes scattering to the smaller landing areas among the flashing lights of the flat-topped business blocks. Among them Smithy threaded his way toward the green-lighted transfer zone, where he spiraled upward.

At ten thousand he was on his course. He set the gyro-control which would fly the ship more surely than any human hands, and the air-speed indicator crept up to the four hundred and fifty miles an hour that Culver had promised. Not till then did he give the man in the forward cockpit the details of his "nightmare."

He had not finished answering the other's incredulous questions when he throttled down to slow cruising speed and nosed the ship toward a distant expanse of sage-blurred sand.

Outside the restricted metropolitan area he had already dropped out of the chill wind that struck them at ten thousand. Behind them and off to the right was the gray rampart of the Sierra. Ahead a rough circle of darker hills enclosed the great bowl he had learned to know as Tonah Basin.

Some feeling of unreality in his own experiences must have crept into his mind; unconsciously he had been questioning his own sanity. Now, at sight of the sandy waste where he and Rawson had labored, with the dark slopes of desolate craters looming ahead and a blot of burned wreckage directly below to mark the site of their camp, the horrible reality of it gripped him again.

He could not speak at first. The air of the five-thousand level was not uncomfortably warm, but Smithy was feeling again the baking heat of that desert land; again he was with Rawson in the volcanic crater; Dean was calling to him, warning him....

A sharp question from Culver was repeated twice before Smithy could reply.

He side-slipped in above the crater's ragged rim, heedless of down-drafts — the power of the DeGrosse motor would pull them out of anything in a ten-thousand-foot vertical climb if need arose. Smithy was pointing toward a confusion of shining black rock.

"Over there," he told Culver. Then he was shouting into the telephone transmitter. "It's open," he said. "That's where Dean went down — and there they are! Look, man, there — there!"

CHAPTER XIV
EMERGENCY ORDER

The throat of the old volcano was a pit of blackness in the midst of gray ash and the red-yellow of cinders. Beside it were other flecks of color: red, moving bodies; metal, that twinkled brightly under the desert sun—and in an instant they were gone. Nor did Smithy, throwing the thundering plane close over that place, know how near he had passed to sudden, invisible death. Rugged pinnacles of rock were ahead. The plane under Smithy's hands vaulted over them and roared on above the desert.

"Did you see them?" Smithy was shouting.

The man in the forward cockpit turned to face his pilot. "I am apologizing, Smith, for all the things I have been thinking and haven't said. We've got a job on our hands. Now let's find that fool sheriff who thinks he's hunting for drunken Indians. We must warn him."

Smithy wondered at the wisps of blue smoke still rising from the ruins of Seven Palms as he drove in above it. It seemed years since he had left the Basin, yet the wreckage of this little town, only five miles outside, still smoldered.

Colonel Culver was shouting to him. "East," he said. "Swing east. There's fighting over there." Then, in his usual cool tone: "I'll take the ship, Smith. Give then a burst or two from up here—perhaps the sheriff can use a little help."

Across the yellow sand ran a desert road. Ten miles away black smoke clouds were lifting. Smithy knew there had been a little settlement there. A dozen houses, perhaps, and a gasoline station. At half that distance the clear sunlight showed moving objects on the sand: automobiles, smaller dots that were running them. They came suddenly to sharp visibility as the plane drew near. Tiny bursts of white meant rifle fire.

They were a thousand feet up and close when Smithy saw the first car vanish in flame. Others followed swiftly. Men were falling. A dozen of them had made up the sheriff's posse, and now, like the cars, they, too, burst into flame and either vanished utterly or, like living torches, were cast down upon the sand.

Still no sign of the enemy, more than the ripping stab of green fire from a sand dune at one side. They were over and past before Smithy, looking back, saw the red ones leap out into view.

Culver must have seen them in the same instant. He throttled down to a safe banking speed. Opened full, the DeGrosse would have whipped them around in a turn that would have meant instant death. From five miles distant they shot in on a long slant. Smithy's hands were off the stick. It was Culver's ship now.

He saw the man peering through his sights, then the roar of the motor held other, sharper sounds. Thin flames were stabbing through the propeller disk, and he knew that the bow guns were sending messengers on ahead where red figures waited on the sand.

Their trajectory flattened. Culver half rolled the ship as they sped overhead. "He wants a look at them," Smithy was thinking. Then a blast of heat struck him full in the face.

It was Smithy's hand on the stick that righted the ship; only the instant response of the big DeGrosse motor tore them up and away from the sands that were reaching for those wings.

His face was seared, but the pain of it was forgotten in the knowledge that their drunken, twisting flight had whipped out the fire licking back from the forward cockpit. He saw Culver's head, fallen awkwardly to one side. The helmet in one part was charred to a crisp.

He leveled off. He was thinking: "Another man gone! Can't I ever fight back? If I only had a gun!" Then he knew he was looking at the pistol grip, where Colonel Culver's brown hand had brought an awkward weapon to life. His lips twisted to a whimsical smile, though his eyes still held the same cold fury, as he whispered: "And I don't even know that the damn thing's loaded—but I'm going to find out!"

They were clustered on the sands below him as he roared overhead. He was flying at two thousand, the throttle open full. Beside the ship a gun swung its long barrel downward. It sputtered almost soundlessly—but where it passed, the sand rose up in spouting fountains.

But his wild speed made the gunfire almost useless. The shell-bursts were spaced too far apart; they straddled the blot of figures.

He came back at five thousand feet, slowly—until the ship lurched, and he saw the right wing tip vanish in a shower of molten metal. He threw the ship over and away from the invisible beam; the plane writhed and twisted across the last half mile of sky. He was over them when he pulled into a tight spiral, then he swung the pistol grip that controlled the gun until the dot in the crystal was merged with the target of clustering red forms. The gun sputtered.

Below the plane, the quiet desert heaved its smooth surface convulsively into the air. Even above the roar of the motor Smithy heard the terrific thunder of that one long explosion.

Above the rim of the forward cockpit Culver's head rolled uneasily; his voice, thick and uncertain, came back through the phone; and later — only a matter of minutes later, though fifty miles away — Smithy set the plane down on a level expanse of sand and tore frantically at his belt. Colonel Culver was weakly raising his head.

"What hit us?" he demanded when Smithy got to him. "Did I crash?" He looked about him with dazed eyes from which he never would have seen again, but for the protection of his goggles.

"Fire," said Smithy tersely. "They did it, the devils, and it wasn't a flame-thrower, either. There wasn't a flash of their cursed green light. It just flicked us for a second. You got the worst of it. Your half roll saved us. That thing, whatever it was, would have ripped our left wing off in a second."

He was looking at the forward cockpit where the metal fuselage was melted. The leather cushioning around the edge was black and charred. Culver's helmet had protected him, but half of his face was seared as if it had been struck by a white flame.

"But we got some of them: they know we can hit back...." Smithy began, but knew he was speaking to deaf ears. Again his passenger had lapsed into unconsciousness.

Quickly he disconnected their own radio receiver and threw on the emergency radio siren. Ahead of them for a hundred miles an invisible beam was carrying the discordant blast. Then, with throttle open full, regardless of levels and of air traffic that tore frenziedly from his path, he drove straight for the home field.

In the office of the Governor, the radio newscaster was announcing last-minute items of interest. The Governor switched off the instrument as Smithy entered, supporting the tall figure of Colonel Culver, whose face and head were swathed in bandages. Culver had insisted upon accompanying him for the rendering of their report, though Smithy had to do the talking for both of them.

He outlined their experience in brief sentences. "And now," he was saying grimly, "you can go as far as you please, Governor. You've got a man's sized fight on your hands. We don't know how many there are of them. We don't know how fast they'll spread out, but—"

A shrill wail interrupted him. From the newscasting instrument came a flash of red that filled the room. The crystal, the emergency call, installed on all radios within the past year and never yet used, was clamoring for the country's attention.

Governor Drake sprang to switch it on, and tried to explain to Smithy as he did so. "It's out of my hands now," he said. "Washington has—" Then the radio came on with a voice which shouted:

"Emergency order. All aircraft take notice. Mole-men"—Smithy started at the sound of the word; it was the name he had given them himself—"Mole-men are invading Western states. A new race. They have come from within the earth. In Arizona, three ships of the Transcontinental Day Line, Southern Division, have been destroyed with the loss of all passengers and crew. Shattered in air.

"It is war, war with an unknown race. Goldfield, Nevada, is in ruins. Heavy loss of life. Federal Government taking control. Air-Control Board orders traffic to avoid following areas...."

There followed a list of locations, while still the red crystal blazed its warning across the land and to all aircraft in the skies. Southern California, Arizona, Nevada—Southern Transcontinental Routes closed; all except military aircraft grounded in restricted areas.

Smithy's excitement had left him. In his mind he was looking far off, deep under the surface of the world. "They've been there," he said quietly, "thousands of years. A new race—and they've just now learned of this other world outside. Three ships downed! They picked them off in the air just as they tried to do with us. I knew we had a fight on our hands."

His voice died to silence in the room where now the new announcer was giving a list of the dead—a room where men were speechless before an emergency no man could have foreseen. But Smithy's eyes, gazing far off, saw nothing of that room. Again he was seated on an outthrust point of rock, Dean Rawson beside him, and from the black depths beneath a man's voice was rising clearly, mockingly it seemed, in song:

> "You're pokin' through the crust of hell
> And braggin' too damn loud of it,
> For, when you get to hell, you'll find
> The devil there to pay!"

"The devil is there to pay," Smithy repeated softly. He leaned across and placed one hand on Colonel Culver's knee. "With your assistance, Colonel, I'd like to go down there and find him. You and I, we know the way—we'll organize an expedition. Maybe we can settle that debt."

CHAPTER XV
THE LAKE OF FIRE

Before a barrier of gold, waist-high, Dean Rawson stood tense and rigid. Behind him the great cave-room swarmed with warriors, leaders, doubtless, of the unholy hordes. But beyond the barrier were the real leaders of the Mole-men tribes — Phee-e-al, ruler in chief, and his clustering guard of high priests. In the flooding light from the wall, their eyes were circles of dead-white skin. A black speck glinted wickedly in the center of each.

Phee-e-al was speaking. His artificially whitened face grimaced hideously; the shrill whistling voice made no comprehensible sound. But in some manner Rawson gathered a dim realization of what his gestures meant.

Phee-e-al pointed at the captive; and one lean hand, with talons more suggestive of a bird of prey than of a human hand, pointed downward. "Gevarro," he said. The word was repeated many times in the course of his whistling talk.

"Gevarro" — what did it mean? Then Rawson remembered. It was the word he had heard in his dreams, the name of the lake of fire.

The voices of the priests rose in a shrill chorus of protests, and even Phee-e-al stood silent. They crowded about their ruler, and Rawson knew they were demanding him for themselves. Then the one who still held a human body in his arms sprang forward and his long talons worked unspeakable mutilation upon the body and face.

Rawson averted his eyes from the ghastly spectacle. For, swiftly, he was seeing something more horrifying than this desecration of a dead body; he was seeing himself, still living, tortured and torn by those same beastly hands. The dead face of Sheriff Downer was staring at him from red, eyeless sockets as with one leap Rawson threw himself over the golden wall. Ten leaping strides away was his gun. In that instant of realization, he knew why his life had been spared.

In the room of fire he had destroyed their priest. They had saved him for further torture.

To get his hands on the gun, to die fighting — the thought was an unspoken prayer in his mind. Behind him the room echoed with demoniac shrieks. Before him was the metal stand. His outstretched hands fell just short of the blue .45 as he crashed to the floor. The copper ones were upon him.

Half stunned by the fall, he hardly knew when they dragged him to his feet. He was facing the golden figure of Phee-e-al, but now the ruler's indecision had vanished. He was exercising his full authority and even Rawson's throbbing brain comprehended the doom that was being pronounced.

"Gevarro!" he was shrieking. "Gevarro!"

Beside him a priest swept the metal table clear. Rawson's clothing, the gun, the radio receiver, all were snatched up and hurled into one of the massive chests. Phee-e-al was still shouting shrill commands. An instant later Rawson was lifted in air, rushed to the barrier and thrown bodily from the sacred premises he had invaded. Then the hands of the red guard closed about him before he could struggle to his feet. A shining object swung down above his head. It was the last he knew.

His dreams were of falling. Always when he half roused to consciousness he was aware of that smooth, even descent, and he knew it had continued for hours.

Once he saw black walls slipping smoothly past, upward, always upward. Gropingly he tried to marshal his facts into some understandable sequence. He was falling, falling toward the center of the earth, and this that he saw was not rock, or any metal such as he knew.

"It's all different," he told himself dully, "new kind of matter. Rock would flow; this stands the pressure." But he knew the air pressure had built up tremendously. The blood was pounding in his ears. He wanted to sleep.

It was the heat that awakened him. The air was stifling him, suffocating. He was struggling to move his heavy body, fighting against this nightmare of heat when he opened his eyes and knew that he was in a place of light. First to be seen were walls, no longer black, no longer even with the characteristics of rock, or even metal. Here, as Rawson had sensed, was new material to form the core of a world. It would have been red in an ordinary light. It was transformed to orange, strangely terrifying in the blazing flood of yellow brilliance that came from the tunnel's end.

Rawson's brain was not working clearly. An unendurable weight seemed pressing upon him—the air pressure, he thought, to which he had not yet become accustomed. And the air, itself, hot—hot!

A breeze blew steadily past toward that place of yellow horror at the tunnel's end. Yellow, that reflected light; but its source was a searing, dazzling white in the one brief instant when Rawson dared turn his eyes.

Hands held him erect, red, gripping hands. One, whose body seemed molten copper in that fierce glare, approached. His hand described a circle over Rawson's bare chest. Straight lines radiated out from the circle, lines of stabbing pain for the helpless man. He had seen the same emblem in the temple of fire, again in the big room where Phee-e-al had stood.

The living sacrifice was prepared. Burned into his bare flesh was the emblem of their legendary sun-god. The priests, their bodies coated with a flashing coppery film that must somehow be heat-resistant, had him in their grasp.

The red warriors had fallen back. Then Phee-e-al appeared; he joined the march of death of which Dean Rawson formed the head. Voices were chanting—somewhere a trumpet blared. Then Rawson, moving like one in a dream, knew the priests were guiding him toward that waiting, incredible heat.

The tunnel's end was near. About him was an inferno where heat and hot colors blended. The whole world seemed aflame, but beyond the tunnel's end was a seething pit upon which no human eyes could look and live.

One glimpse only of the unbearable whiteness beneath which was the lake of fire, then the chains of his stupor broke and Dean Rawson struggled frenziedly in the grip of two copper giants.

They had been chanting a shrill monotonous refrain. They ceased now as they fought to throw the man out past that last ten paces where even they dared not go.

Rawson was beyond conscious thought. Eyes closed against the unendurable heat, he fought blindly, desperately, then knew his last strength was going from him. Still struggling he opened his eyes; some thought of meeting death face to face compelled him.

A hideous coppery face glared close into his own. Miraculously it vanished, disappeared in a cloud of white. Then the blazing walls were gone—there was nothing in all the world but rushing clouds of whiteness, shrieking winds, the roar of an explosion—and cold, so biting that it burned like heat.

Vaguely he wondered at the hands that still clutched at him. Dimly he sensed other bodies close to his, other hands that tore him free where he lay, still struggling with the priests, upon the floor. A narrow opening was in the wall, a blur of darkness in the billowing white clouds. They were dragging him into it, those others who held him, and they were white—white as the vapor that whirled about him.

Ahead, the girl of his former dreams was guiding him, her hand cool and soft in his. Others helped him; he ran stumblingly where they led down a steep and narrow way.

The White Ones! In a vision they had reached out to him before. Was this, too, a dream? Was it only the delirium of death? That burst of cold—had it truly been liquid fires, wrapping him around?

Dean Rawson could not be sure. He knew only that his fate lay wholly in the hands of these White Ones—and that hideous eyes in the coppery face of a priest had glared at them as they fled.

CHAPTER XVI
THE METAL SHELL

Dean Rawson had passed through a nerve-racking experience. It was not a question of courage—Rawson had plenty of that—but there are times when a man's nervous system is shocked almost to insensibility by sheer horror. Not at once did he realize what was happening.

The Voice of the Mountain heralds Rawson's Messianic coming to the White Ones in their hour of need.

Perhaps it was the sound of pursuit that jarred him out of the fog clouding all his thoughts and perceptions. It was like the sound of fighting animals—cat-beasts—whose snarls had risen to screaming, squalling shrieks of rage. It was sheer beastliness, the din that echoed through that narrow passage.

Ahead of him the girl was running. She held a light in her hand. Soft wrappings of cloth hung loosely from her waist; like her golden hair, it was flung backward in the strong draft of air against which they were struggling. She was outlined clearly before the red, rock-like masses where her light was falling; she was running swiftly, gracefully, like a wild, woodland nymph.

Two men, their milk-white bodies naked but for the thick folds of their loin cloths, were beside Rawson, helping him along. Two others followed. And, by their haste and their odd whispered words of alarm, he knew that pursuit had not been expected; they must have thought to get away unobserved.

Rawson felt his strength returning. He shook himself free from those who tried to aid him. He was amazed at how easily he ran: his weight was a mere nothing; his efforts were expended in driving his body against the blast of wind. The air seemed dense, thick; he had almost the feeling of forcing himself through water.

Ahead of him the girl darted abruptly through a narrow crack in the wall. Rawson followed—and then began a wild race through a network of connecting passages, a vast labyrinth of caves, more like fractures in this strange red substance which Rawson could think of only as rock, for lack of a more accurate name, until at last there was no sound except that of their own hurrying feet.

They stopped and stood panting in one of the wider passages. He heard nothing but the endless rush of the wind. For the first time Rawson became aware of his own almost naked condition.

The mole-men had prepared him for the sacrifice. They had decked him with a loin cloth of woven gold. It felt cold to the touch, and Rawson did not doubt its being made of fine threads of the precious metal. About his neck hung a gold chain with a heavy object suspended; he tore it off, and found again a representation of a golden sun. The copper priests had arrayed him to meet their fire-god, and again Rawson wondered at the emblem they employed.

"What in the name of the starlit heavens," he demanded silently of himself, "could this buried race know of the sun?"

The others were watching him. In the glow of that strange light held by the girl he saw them smiling. They were congratulating one another with odd, soft-syllabled words. And Rawson, ignorant of their tongue, was mute, when his whole soul cried out to thank them.

He gripped the hands of the men. They were as tall as himself, their gaze level with his own. Their faces were human, friendly; their eyes sparkled and smiled into his. Then he turned to the girl.

She had seen the method of greeting this stranger employed. She extended her hand—a white hand, slim, soft, cool. And Rawson, choking with emotion, knowing that here was the one who had first seen him and who had returned to save him, a stranger, bent low above that hand, held in his own so rough and burned, and pressed his lips to the slender fingers in a quick caress.

When he raised his head she was looking at him oddly; her eyes were deep, serious and unsmiling. He wondered if, blunderingly, he had offended her. He could not know; he did not know their customs.

Again the slim girlish figure turned; her jeweled breast-plates flashed as she led the others on where always the way led upward and the wind pressed against them unceasingly.

The White Ones wore sandals that seemed woven of glass. Rawson's bare feet were bruised and sore, for those narrower clefts had been paved only with broken fragments of the red walls. He moved less easily now. The heavy, beating air tired him; the lightness of his body made it all the more difficult to fight the steady wind. Still he followed the white figure of the girl where her light was flashing on endless walls of red.

In his ears a new sound was registering. Above the rush of the air, that now was soft and warm, a new note had risen to a hollow, unremitting roar. He knew that for some time he had been hearing it faintly. It grew louder, one long, steady, unchanging note, as they advanced. It was a deafening reverberation that seemed shaking the whole earth when they came at last to an open room.

It beat upon him thunderously. As deep as the deepest tone of a mighty organ, like a thousand gigantic organs welded in one, it roared and shook him through and through with its single note.

Exhausted by his wild flight, surrounded by this maelstrom of sound, he sank to the floor and let his laboring lungs have their way. But his eyes were searching the big room.

The great cave was too regularly formed to have had a natural origin. The light that the girl had carried gave only feeble illumination in so great a space that had so evidently been hollowed out of the solid red matter.

The light flashed here and there as the girl and her companions moved away. They were circling the room. Rawson saw the irregular outlines of entrances to many dark passages like the one through which they had come. The red rock-mass seemingly had been riven and torn, and apparently in front of each opening the white figures fought against the rush of outgoing air. Rawson felt the same current sweeping and whirling gustily about him.

Now his companions were across the room, and between him and them in the center of the floor he saw the mouth of a black well, a pit some twenty or more feet across. Directly above, where the red rock stuff formed a domed ceiling, he found a counterpart of the pit below—another great bore or open shaft, roughly circular. Apparently it went straight on up and was a continuation of that lower pit.

"This room was cut out," Rawson was thinking, "by the white people or the mole-men—Lord knows who, or when, or why. Cut out around this big shaft...."

His thoughts trailed off. Even thinking seemed impossible under the battering of the roaring noise that pounded about him. Then another thought pierced through the bedlam. He had found the source of the uproar.

That upper shaft, the hole that went on up, must be plugged. There was no outlet that way, and this air that drove endlessly upward from the room must be coming from the lower shaft. It was striking up into that upper cavity.

An organ pipe, truly. But whence came the unending blast of air to keep that gigantic instrument in operation? Rawson dropped to his knees

and crept slowly across the floor toward the pit. He must test his theory—
see if that was where the air was driving in.

Just short of the brink he stopped. The girl had called—a cry of alarm.
She was running swiftly toward him, circling the pit. And Rawson, as she
tugged at him, trying to draw him back, knew that she had mistaken his
motive. She had thought he was going to cast himself down.

He did not need to go farther. He was close to the edge. And now,
even above that roaring sound he heard the rush of the column of air. He
seated himself on the stone floor and smiled up at the girl reassuringly.
Her eyes that had been dark with fear changed swiftly to a look so sweetly,
beautifully tender that Dean Rawson found himself thrilled and shaken by
an emotion that set his nerves to quivering even more than did the sonorous
vibration from above.

Her companions had joined her. Dean saw her eyes regarding them
steadily. Then, as if reaching some sudden final conclusion in her own
mind, she dropped swiftly to her knees beside him, raised one of his hands
in hers and pressed her soft lips against it.

And Dean, even had he known their language, could not in that
moment have spoken. There had been something in the look of her eyes and
the soft touch of her lips that of themselves went far beyond words.

"You darling," he was whispering softly to himself as the girl sprang to
her feet and walked swiftly away, the others following.

"An angel, no less—down in this damned place!"

He wondered, as he watched the flickering light far across the room,
what destination they could be bound for. Surely no one so radiantly
beautiful could inhabit a world of endless dungeons like that where the
mole-men lived. But if not that, then what? Where would their next journey
take them? And in what direction would they go?

Again Rawson's thoughts were submerged beneath his own weariness.
This air that beat about him had seemed cool after the terrific heat that drove
in off the Lake of Fire. Now he realized that the air itself was hot. His one
spurt of strength and energy had been expended.

He watched the men disappear into one of the passages, but he roused
himself when they returned. They were clinging to a strange device, a metal
cylinder that floated in air above their heads like a dirigible on end. It was
about eight feet in diameter and some fourteen feet in height; both upper
and lower ends were rounded. A cage of parallel bars enclosed it from end
to end; like springs of steel they extended from top to bottom where they
curved in and were attached to the rounded ends.

Rawson sat up quickly and stared in startled amazement at the thing glinting like polished aluminum in the light. And his engineer's mind responded as much to that smooth finish and the evident workmanship that had entered into the making of this thing as it did to the object itself.

The girl placed her light on the floor. She, too, reached up and gripped a bar of the protecting cage to which the others were holding. With her added weight and strength they drew it down almost to the floor. Rawson knew by their efforts that they were dealing with something actually buoyant, a metal balloon. One of the men, still putting his weight on the bars, reached in and opened a door in the smooth shell. He stepped inside, and a moment later the big shell dropped to the floor and, still vertical, stood on the lower rounded end of the protecting cage, rocking gently as the hot whirling wind hit it.

They were communicating among themselves by signs. Rawson saw them motioning. Speech was useless in that roaring, pandemonium-filled room.

She was motioning for him to follow. One of the men circled that central pit, came beside Rawson and helped him to his feet, steadying him as they crossed the room. The girl had entered the big metal shell. Dean saw the glow of her torch shining through the open doorway and through two other windows of crystal glass.

The big room had grown dimmer. The high ceiling was lost in murky shadows. All the room was dark save where that light struck upon walls and floor to make them glow blood-red. The waiting lighted shell seemed a haven of refuge. To get inside, close the door, lock out some of this unendurable, battering sound — it was all Rawson asked, all he could think.

The door closed. He was within the shell, standing on a smooth metal floor. The others were beside him. Dully he wondered what wild adventure was ahead.

He had expected — he hardly knew what. But there should have been machinery of some sort. If this weird balloon thing was actually to carry them, there must be some mechanism, some propelling power. And instead he saw nothing but the shining walls of the circular room and at the exact center, reaching from floor to ceiling, a six-inch metal post that thickened to a boxlike form on a level with his eyes. There was a plate on the side of that box, a cover, and clamps that held it in place, and on an adjoining side two little levers, one near the top of the box, the other near the bottom.

His one all-inclusive glance showed him bull's-eye windows in the ceiling. There were more of them in the floor. One curved bar, circling the room, was mounted on brackets against the wall. They were telling him by signs that he was to put his hands on it and hang on. One of the men was beside that central post. He too gripped at a projecting hand-hold. His other hand was on the lower lever.

Rawson knew his disappointment was unreasonable, but his weary mind was tired of mysteries. Some understandable bit of machinery would have been reassuring. And then in his next thought he asked himself what difference did it make. If this childish balloon thing were really capable of carrying them somewhere, what of it? It could only mean more of this hideous inner world that grew more unbearably fantastic with each new experience.

His life had been saved. True, but for what end? The girl's eyes were upon him, reading the expression on his face. She smiled encouragingly. Then Rawson's hands tightened upon the metal bar. The man who stood by the central post had moved one lever the merest trifle. Rawson felt the floor lifting beneath him. Then the shell, like a bubble of metal, pitched and tossed as the powerful air currents caught it.

His own lightness saved him from injury. He gripped the bar and held himself free of the wall. The round top of their strange craft grated against the domed roof. Then again the ship steadied and seemed motionless, and Rawson knew they had slipped up into the still air of that upper shaft.

For one wild instant, filled with impossible hope, Rawson saw this as a means of ascent to his own world. Then reason tore those wild hopes to shreds.

"It's closed up above," he thought. "It must be. That's why it sounded that way. That's why the air drove off through those side passages."

The next instant held no time for thought. Rawson's whole attention was concentrated upon the bar to which he clung. For, quicker than thought, the metal shell, the little cylindrical world in which he and these others were, fell swiftly beneath them.

His body twisted in mid-air. He knew the others were being thrown in the same manner. Then, what an instant before had been the ceiling was now a floor beneath his feet, pressing up against him and giving him weight — and by the whistling rush of the air that tore past their shell he knew they had fallen with marvelous swiftness straight down through the throat of that lower shaft.

And now what had been down was up. The ceiling of this strange room was now their floor, but Rawson was not deceived. "Acceleration," he said. "It's crowding us. The shell tends to fall faster than we do. It's like an elevator traveling downward at a swifter rate than a free falling body."

He had glimpsed the glassy-side of that well into which he knew they had been flung. He knew that the shrieks that filled the room time and again were caused by the touching of their shell's guiding and protecting bars against one glassy wall. Those sounds came always from the same side and Rawson found momentary satisfaction in his own understanding of the phenomenon.

"We're falling free," he argued within his own mind, "falling toward the center of the earth. And a falling body wouldn't follow a vertical course. It would tend to hug against one wall." And by that he knew something of their speed. The necessity for it was apparent a moment later.

Above his head the bull's-eyes pointing forward in the direction of their flight were faintly red. Swiftly they changed to crimson. Rawson was standing beside a window in the wall of their craft. That, too, grew quickly to an area of dazzling brightness. Slowly the heat struck in. The air in the little room was stifling. He saw the girl turn her head and give a sharp order.

The man by the central post responded with another slight movement of the lever. Beneath Rawson's feet the floor pressed upward in a surge of speed that bent his knees and bore him downward. Under his hands the rod to which he clung was hot. The shining walls were dimly glowing. They were being hurled through the very heart of hell....

And then it was past. The crimson horror beyond those windows grew dull and then black. In the blunt nose of their craft a tiny crevice must have opened. The one who drove that projectile in its shrieking flight had touched another control that Rawson had not before seen. And with a piercing shriek a thin jet of cold air drove down into the hot room.

No wine could have been one-half so potent. That thin jet filled the room with buffeting whirlwinds that grew quickly cold.

Then their speed was checked. Abruptly Rawson was weightless, his body hanging in air, moved only as he moved his hand upon the bar. Only a few feet away was the body of the girl floating weightless like himself. The others were shouting loud words of satisfaction, but her face was turned toward Rawson, her eyes were smiling into his; while, outside the little shell that fell in meteor flight, were only shrieking winds and the blackness into which they plunged.

CHAPTER XVII
GOR

Through an ordinary experience, Dean Rawson, like any other man, would have kept unconscious measurement of the passing time. An hour, no matter how crowded, would still have been an hour that his mind could measure and grasp. But now he had no least idea of the hours or minutes that had marked their flight. Each lagging second was an age in passing. Even the flashing thoughts that drove swiftly through his mind seemed slow and laborious. Painstakingly he marshaled his few facts.

"They know what they're about, that's one thing dead sure. They're onto their job, and they've got something here that beats anything we've ever had." He mentally nailed that one fact down and passed on to the next. "And that's the bow end of our ship, up there." He looked above him at a dented place in the ceiling, the ceiling that had been the floor of the room when first he stepped into it. "There isn't any up or down any more. I've been flipped back and forth every time we slowed down or accelerated until I don't know where I'm at, but I saw that dented plate in the floor when I got in and we started falling in that direction. But whether we're falling toward the center of the earth still or whether we passed the center back there at that hot spot and now this crazy, senseless shell is flying on and up, perhaps these people know—I don't!"

Then fact No. 3. "They live somewhere inside here. They're taking me there, of course. It must mean there's a race of them—and they don't like the mole-men. They know the way back, too, and if they'll help me.... Perhaps the fighting's not over yet!"

Through more endless, age-long seconds there passed through Rawson's mind entrancing visions. An army of men like these White Ones, himself at their head. They were armed with strange weapons; they were invading the mole-men's world....

The girl was reaching toward him. She laid one hand upon his, then pointed overhead.

Rawson looked quickly above. The glowing bull's-eyes startled him, then he knew it was white-light he was seeing, not the red threat of glowing rock. Their speed had been steadily cut down as the air pressure lessened. "They're decompressing," he thought. "They're working slowly into the lesser pressure."

The passing air no longer shrieked insanely. Above its soft rushing sound he heard the girl's voice; it was clear, vibrant with happiness. Her hand closed convulsively over his; her eyes beneath their long lashes smiled unspoken words of welcome, of comradeship, and of something more.

Within their room her light, which at close range seemed only a slender bar of metal with a brilliantly glowing end, had been clamped in a bracket against the wall. The illumination had seemed brilliant, now suddenly it was pale and dim.

Through the bull's-eyes above, a brighter light was shining, clear and golden, like the light of the sun on a brilliant and cloudless day. And to Rawson, who felt that he had spent a lifetime in the gloomy dungeons of that inner world, that flooding brilliance was more than mere light. It was the promise of release, the very essence of hope. His eyes clung to these little round windows; then the larger glass beside him blazed forth with the bright sunlight of an open world that was unbearable to one who had lived so long in darkness.

He held tightly to that slim hand that remained so confidingly within his own.

"It isn't true," Rawson was telling himself frantically. "It can't be true. It must be a delusion, another dream."

He gripped the girl's hand in what must have been a painful clasp. He told himself that she at least was real. Her lovely face was before him when at last he could bear to open his eyes.

About him were the others. The cylinder rested firmly upon a surface of pale-rose quartz. Inside the shell he saw the floor where he had stood, and with that he added one more fact to the few he had gotten together. There was no dent in the floor. The shell's position was reversed. What had been up was now down. Rawson knew he was standing firmly, with what seemed his normal earth weight, upon a smooth surface of rock; he knew that he was standing head down as compared with his position at the beginning of their flight—as compared, too, with the way he had stood in the mole-men's world and in his own world up above.

"I've passed the center of the world." The words were ringing in his brain. And then reason shot in a quick denial. "You're as heavy as you were on earth," he told himself. "You'd have to go through and on to the other side, the opposite surface of the world, before your weight would come back like that!"

"What could it mean?" he was demanding as his eyes came back from the machine and swept around over a gorgeous, glittering panorama of crystal mountains, rose and white. Fields of strange plants, vividly green; a whole world that rioted madly in a luxury of color. Before him the girl stood smiling. Every line of her quivering figure spoke eloquently of her joy in seeing this world through Rawson's eyes.

A man was approaching, a man like the others, yet whose oval face strangely resembled that of the girl. She led Rawson toward him, then Rawson, stopping, jerked backward in uncontrollable amazement, for the tall man drawing near had spoken. His lips were open, moving, and from them came sounds which to Rawson were absolutely unbelievable:

"Stranger," said the newcomer, "in the name of the Holy Mountain, and in the Mountain's language and words, I bid you welcome."

And Rawson, too stunned for coherent thought, could only stammer in what was half a shout: "But you're speaking my language. You're talking the way we talk on earth. Am I crazy? Stark, raving crazy?"

But even the sound of the man's voice could not have prepared him for what followed. There was amazement written on the face of the man. And the girl who stood beside him—her eyes that had been smiling were wide and staring in utter fear. Then she and the man and the other white figures nearby dropped suddenly to kneel humbly before him. Their faces were hidden from him, covered by their hands as they bent their heads low. He heard the man's voice:

"He speaks with the tongue of the Mountain! He comes from the Land of the Sun, from Lah-o-tah, at the top of the world! And I, Gor, am permitted to hear his voice!"

CHAPTER XVIII
THE DANCE OF DEATH

Through an airplane's thick windows of shatter-proof glass, so tough and resilient that a machine-gun bullet would only make a temporary dent, the midday sun flashed brightly as the big ship rolled. Along each side of the small room, high up under the curve of the cabin roof, windows were ranged. Others like them were in the floor. And, above, the same glass made a transparent dome from which an observer could see on all sides.

Outside was the thunderous roar of ten giant motors, but inside the cabin—the fire-control room of a dreadnought of the air—that blast of sound became more a reverberation and a trembling than actual noise.

Certainly the sound of motors and of slashing propellers, as the battle plane roared up into the sky, did not prevent free conversation among the three men in the room. Yet there was neither laughter nor idle talk.

At a built-in desk, before a battery of instruments, sat Farrell, the captain of the ship. Farther aft, in solidly anchored chairs, Colonel Culver and Smithy were seated. Occasionally the captain spoke into a transmitter, cutting in by phone on different stations about the ship.

"Check up on that right-wing gun, Sergeant—number two of the top wing-battery. Recoil mechanism is reported stiff.... Tell Chicago, Lieutenant, we will want one thousand gallons in the air—gas only—no oil needed.... Gun room? Have the gun crews get some sleep. They'll have to stand by later on...."

Colonel Culver spoke musingly. "Guerilla warfare, the hardest kind to meet."

Smithy nodded absently. He rose and stared from one of the side windows that was just level with his eyes. He could see nothing but the broad expanse of wing, a sheet of smooth gray metal. Along its leading edge was a row of shimmering disks where great propellers whirled. From the top of the wing a two-inch Rickert recoilless thrust forth its snout; it rose in air till the whole weapon was visible, then settled again and buried itself inside the wing.

They were testing a gun. Smithy knew that inside that wing section were other guns, and men, and smoothly running motors. The whole ship was only a giant flying wing of which their own central section was merely a thickening.

He looked down through a bull's-eye in the floor. The city they had just left was beneath them. Washington, the nation's capital; the golden dome of the Capitol Building was slipping swiftly astern. Only then did he make a belated reply to Culver's statement.

"Well," he said shortly, "they'll have to meet it their own way. We told them all we knew. And a lot of good that did — not!"

"Five days!" said Culver. "It seems more like five years since the devils first came out. Nobody knows where they will hit next. But they're working north — and there's no trouble in telling where they've been."

Smithy's voice was hot in reply, hot with the intense anger of a young, aggressive man when confronted by the ponderous motion of a big organization getting slowly under way.

"If only we'd gone down underground," he exclaimed; "carried the fight to them! They live there — there must be a whole world underground. We could have carried in power lines, lighting the place as we went along. We could have fought 'em with gas. We'd have paid for it, sure we would, but we'd have given them enough hell to think of down below so they wouldn't raise so much of it up above.

"But no! We had to fight according to the textbooks. And those red devils don't fight that way; they never learned the rules."

"Guerilla warfare," Colonel Culver repeated. "There are certain difficulties about fighting enemies you can't see."

"They're clever," Smithy admitted. "We taught them their lesson down there in the desert — they've never been seen in daylight since. Out at night — and their invisible heat-rays setting fire to a city a mile away, then mopping up with their green flame-throwers if anyone's left. They pick our planes out of the sky even when they're flying without lights. Darkness means nothing to them! It was murder to send troops in against them, troops wiped out to a man! Artillery — that's no good either when we don't know how many of the devils there are, or where they are. There's no profit in shelling the place when the brutes have gone back underground."

Colonel Culver shot a warning glance from Smithy to the seated officer. "About a hundred square miles of the finest fruit country on earth laid waste," he admitted gravely; then sought to turn Smithy from his rebellious mood:

"What's underground, I wonder? Must be a world of caves. Or perhaps these mole-men can follow up a mere crack or a fault line and open it out with their flame-throwers to make a tunnel they can go through."

The plane's captain had caught Culver's glance. "Speak your piece," he said pleasantly. "Don't stop on my account. There's a lot to what Mr. Smith says—but you don't know all that's going on."

He had been half turned. Now he swung about in his little swivel chair, whose base was riveted solidly to the floor and whose safety belt ends dangled as he turned.

"My orders are to deliver you two gentlemen at San Francisco. But there's a show scheduled for to-night down south of there—two hundred planes, big and little, scouts, cruisers, battle planes. They're going to swarm in over when the enemy makes his first crack. There's a devil of a storm in the mountains along the route we would usually take. I'm afraid I'll have to swing off south." He was grinning openly as he turned back to his desk.

Colonel Culver smiled back. "Attaboy!" he said.

But Smithy's forehead was still wrinkled in scowling lines as he walked forward to an adjoining room. "Underground," he was thinking. "We've got to carry the fight to them; got to lick 'em so they'll stay licked. But Rawson—good old Dean—we're too late to help him. And the lives of all the devils left in hell can't pay for that."

Smithy had been dozing. The shrill whistle of a high-pitched siren brought him fully awake in an instant. Culver, too, sprang alertly to his feet. Both men knew the signal was the call to quarters.

They had spread blankets on the floor of the fire-control room. Culver immediately folded his into a compact bundle, and Smithy followed suit, as he said: "That's right; we don't want any feather beds flying around here in case of a mix-up."

Even Culver's simple act of stowing the blankets back in their little compartment thrilled him with what it portended. His nerves were suddenly aquiver with anticipation. A real fight! A determined effort! No telling what these big dreadnoughts could do. Two hundred, big and little, Captain Farrell had said. If they could catch the enemy out in the open, show him up in a blaze of enormous flares....

Captain Farrell was calling them. A section of the floor had been raised up mysteriously to form a platform beneath the shallow dome of the conning tower. Farrell was there, headphones clamped to his ears, one hand on the little switchboard at the base of the glass dome that kept him in touch with every station on the ship. Beside him was the fire-control officer similarly equipped, though his headphone was connected only with the gun crews.

"The enemy's out!" said Captain Farrell. "And not just where they were expected—they're raising fourteen kinds of hell. The ships have been ordered in. I'm hooked up with the radio room now. They're less than a hundred miles ahead. Of course we won't mix in on it, but I thought it best to have my men standing by."

He pressed a little lever on his switchboard and spoke into the mouthpiece of his head-set. "Pilot room? Our two passengers, Colonel Culver and Mr. Smith, are coming forward. Let them see whatever they can of the show."

He gave the two a quick smile and a nod and waved them forward with the binoculars in his free hand. "It will be 'lights out' after you get there. We'll be flying dark except for wing and tail lights up on top. The enemy's movements are uncertain; perhaps he can see us anyway, but we won't advertise ourselves to him."

The ship's bow was a blunt, rounded nose of glass, cut by cross bars of aluminum alloy. That deeper central portion of the big flying wing was carried ten feet forward; it was but one of many details that Smithy had looked at with interest when he had seen the ship waiting for them on the field.

The pilot room was dark when they entered. Only the glow from the instrument panel showed the two men who were seated behind the wheel controls. One of them turned and nodded a welcome.

"Can't offer you gentlemen seats," he said, "but if you'll stand right here behind us you can see the whole works." He did not wait for a reply, but turned back toward the black night ahead.

Smithy glanced past him at the lighted instruments and found the altimeter. Twelve thousand—yes, there was nasty country hereabouts. Then he, too, stared out into the dark at the sky sprinkled with stars, at the vague blur of an unlighted world far below, and off at either side and behind them the quivering lines of cold light where starlight was reflected dimly from the spinning propellers.

Other wing lights winked out as he watched, and he knew that from that moment on, they were invisible from below—invisible to human eyes at least—that they were sweeping on through the darkness like some gargantuan night bird pursuing its prey.

"Flares ahead, sir," one of the pilots had spoken into the mouthpiece of his telephone, spoken lightly, reporting back to Captain Farrell. The words whipped Smithy's head about, and he, too, saw on a distant horizon, the beginning of a white glare.

They were fighting there—two hundred planes roaring downward, one formation following another. In his mind he was seeing it so plainly.

The white blaze of light dead ahead grew broader. It had not been as far distant as he had first thought, and the scene that he had pictured came swiftly to reality.

Their own ship was still at the twelve-thousand-foot level. Ahead, and five thousand feet below, tiny lights, red and white and green, lights whose swift motion made their hundreds seem like thousands instead, were weaving intricate patterns in the night. The flying lights of the fighting planes were on for the planes' own protection; and, too, no further concealment was possible in the glare that shone upward from below.

Settling downward were balls of blinding fire, flares dropped by the squadron of scout planes that had torn through in advance. They lighted brilliantly a valley which, a few hours before, had been one of many like it—square fields, dark green with the foliage of fruit trees, straight lines of crossing roads, houses, and off in the distance a little city.

And now the valley was an inferno of spouting flame. That city was a vast, roaring furnace under smoke clouds of mingled blood-red and black. The valley floor was a place of desolation, of drifting smoke and of flashing shell-bursts as the fleet swept in above.

The myriad lights of the planes had drawn into a circle, a great whirlpool of lines that revolved above a mile-wide section of that valley.

Beside Smithy a wheel control was moving. He clung to the pilot's seat as their own plane banked and nosed downward. And now he shouted aloud to Culver:

"The mole-men! There they are! Thousands of them!"

He was pointing between the two pilots as their own plane swept down. He could see them plainly now, clotted masses of dark figures surging frenziedly to and fro. For an instant he saw them—then that part of the world where they had been was a seething inferno of bursting bombs and shells.

Beside him Colonel Culver spoke quietly: "Caught them cold! That's handing it to them."

Their own plane had leveled off. With motors throttled they were drifting slowly past, only a thousand feet higher than the circling planes just off at one side. Culver's quiet tones rose to a hoarse shout: "The ships! My God, they're falling!"

His wild cry ended in a gasp. Beside him Smithy, in breathless horror, like Culver, was staring at that whirlpool of tiny lights that had gone suddenly from smooth circular motion into frenzied confusion, or vanished in the yellow glare of exploding gas tanks. The light of their own white flares picked them out in ghastly clarity as they fell.

Straight, vertical lines of yellow were burning planes. Again they made horrible zigzag darts and flashed down into view torn and helpless, while others, tens and scores of others with crumpled wings, joined the mad dance of death.

Smithy knew that he could never tear his eyes away from the sight. Yet within him something was clamoring for his attention. "They didn't do it from below!" that something was shouting. "Not down in that hell. There are more of them somewhere." Then somehow, he forced his eyes to stare ahead and outside of that circle of fearful fascination and he knew that for an instant he was seeing a single stab of green flame.

One single light on the darkness of a little knoll that stood close beside this place of white flame and destruction. One light — and in the valley there had flashed a million brighter. It had shone but an instant, but, to Smithy, watching, it was the same he had seen when their own camp was attacked. And now it was Smithy who was abruptly stone cold.

One hand closed upon a pilot's shoulder with a grip of steel; his other pointed. "Down there — they're hiding back of that hill, picking off our ships from the side." And then, like a guiding beacon, a point of green showed once more.

The plane banked sharply while one of the pilots spoke crisp, clearly enunciated words into his phone. He listened; then: "Right!" he snapped.

"Power dive for bow-gun firing. Level off for bombing from five hundred feet."

Off into the night they were headed. Then a left bank and turn brought the place of blazing flares and falling planes swinging smoothly into view; they were flying toward it.

Against the white glare in the valley of death was a hill, roundly outlined. Then the ship's nose sank heavily down; and, from each broad wing, in straight, forward-stabbing lines, was the steady lightning of the Rickert batteries in action.

The pilot's room was a place of unbearable sound. The crash of gunfire, it seemed, must crush the glass wall like an eggshell by the sheer impact of its own thunder. In that pandemonium Smithy never knew when they flattened out. He knew only that the hill ahead twinkled brilliantly, and that each flashing light was an exploding shell. He knew when the hill passed beneath them.

Then, in the night, close beside them and just outside the pilot-room glass, was a quick glow of red. The plane lurched and staggered. Smithy clung desperately to the seat ahead. The pilot was fighting madly with the wheel. The roar of bombs from astern, where the bombers had launched their missiles at the approaching hill, was unheard. In a world suddenly gone chaotic he could hear nothing. He knew only that the valley dead ahead was whirling dizzily — that it sank suddenly from sight.

They were crashing. That red glow — they had been hit. Then something hard and firm was pressing against him, pressing irresistibly. It was the last conscious impression upon Smithy's mind.

CHAPTER XIX
THE VOICE OF THE MOUNTAIN

In a strange new world surrounded by a group of kneeling figures of whom one, who called himself Gor, had spoken in Rawson's own tongue, Dean Rawson stood silent. It was all too overwhelming. He could not bring words together to formulate a reply. He only stood and stared with wondering eyes at the exquisite beauty of the world about him, a world flooded with a golden light, faintly tinged with green. Then he looked above him to see the source of that light and found the sun.

Not the sun that he had known, but a flaming ball nevertheless. Straight above it hung, in the center of the heavens, a gleaming disk of pale-green gold, magnificently brilliant. He saw it through lids half closed against its glare. Then his gaze swept back down the blue vault of the heavens, back to a world of impossible beauty.

Directly ahead was a land of desolation, radiant in its barrenness. For every rock, every foot of ground, was made of crystal. Nearby hills were visions of loveliness where the colors of a million rainbows quivered and flashed. Veins of metal showed the rich blues and greens of peacock coloring. Others were scarlet, topaz, green, and all of them took the strange sunlight that flooded them and threw it back in blendings radiant and delicate.

The little hills began a short distance off, two low ranges running directly away. One on either side, they made brilliant walls for the flat valley between, whose foreground was barren rock of rose and white. But beyond the glistening barren stretch were green fields of luxuriant vegetation and in the distance, nestled in the green were clustered masses that might have been a city of men. Still farther on, a single mountain peak, white beyond belief, reared its graceful sweeping sides to a shining apex against the heavens of clear blue.

Slowly Rawson turned. A hundred yards away, at his left, there was water, a sea whose smooth rollers might have been undulating liquid emeralds that broke to infinite flashing gems upon the shore. He swung sharply to the right and found the same expanse of water, perhaps the same distance away.

Then he turned toward the shell, which had been behind him and the shaft from which it had emerged, and into which the air was driving with a ceaseless rushing sound. Now, looking beyond them, he found the same ocean; he was standing on a blunt point of rock projecting into the sea. The rest of this world was one vast expanse of water.

Suddenly Rawson knew that it was unlike any ocean of earth. Instead of finishing on a sharply-cut horizon, that sea of emerald green reached out and still out, and *up*! It did not fall away. It curved upward, until it lost itself in the distance and merged with the blue of the sky. It was the same on all sides.

He swung slowly back to face the land that perhaps was only an island. The kneeling ones had raised their bowed heads. They were regarding him from shining, expectant eyes. Only the girl kept her face averted. Rawson spoke to none of them; the exclamations that his amazement and dismay wrung from his lips were meant for himself.

"It's concave! It curves upward! I'm on the inside of the world! And that sun is the center! But what holds us here? What keeps us from falling?" He passed one hand heavily across his eyes. The excitement of the moment had lifted him above the weariness of muscle and mind. Now fatigue claimed him.

"Sleep," he said dully. "I've got to sleep. I've got to. I'm all in."

Gor was beside him in an instant. "Whatever you wish is yours," he promised.

Rawson was to remember little of that journey toward the habitations of this people. Gor had spoken at times along the way: "... the Land of the Central Sun.... The People of the Light, peaceful and happy in our little world...."

Rawson had roused himself to ask: "Who it at the head of it? Who is the king, the ruler?"

And the tall man beside him had answered humbly: "Always since the beginning one named Gor has led. My father, and those who came before him; now it is I. And when I have gone, my little son will take the name of Gor."

He had glanced toward the girl and his voice had dropped into the soft, liquid syllables of their own tongue. She had smiled back at Gor, though her eyes persistently refused to meet those of Rawson.

Again Gor spoke in words that Rawson could understand.

"I think at times," he said, "it is my daughter Loah, my little Loah-San who really rules. I, knowing not who you were, did not approve of this expedition, but Loah insisted. She had seen you, and—" A glance from the girl cut him short.

The words lingered in Rawson's mind when he awoke. The horrible experience of the past days were no longer predominant. Even his own world seemed of a dim and distant past.

He awoke refreshed. He was in a new world and, for the moment, he asked nothing except to explore its mystery. He bathed under a fountain in an adjoining room, and grinned broadly as he wrapped the folds of the long golden loin cloth about him.

"As well be dead as out of style," he quoted. "And now to find Gor and Loah, and see what the devil all this is about—a talking mountain and a buried race that speaks first-rate American."

Gor was waiting for him in a room whose translucent walls admitted a subdued glow from outside. There was food on a table, strange fruits, and a clear scarlet liquid in a crystal glass. Rawson ate ravenously, then followed Gor.

Outside were houses, whose timbered frames of jet-black contrasted startlingly with the quartz walls they enclosed. The street was thronged with people who drew back to let them pass, and who dropped to their knees in humble worship. Like Gor, the men wore only the loin cloth, but for this gala day, that simple apparel added a note of flashing color. The long cloths wrapped about their hips, and brought up and about the waist where the ends hung free, were brilliant with countless variations of crimson and blue and gold. The same rainbow hues were found in the loose folded cloths that draped themselves like short skirts from the women's waists. Here and there, in the sea of white bodies and scintillant jeweled breast-plates, was one with an additional flash of color, where brilliant silken scarves had been thrown about the shoulders of the younger girls.

"From all the land," said Gor, "they have come to do you honor."

Hardly more than a village, this cluster of strangely beautiful shelters for the People of the Light. Beyond, Rawson saw the country, pastures where animals, weird and strange, were cropping the grass so vividly green; fields of growing things; little crystal houses like fanciful, glistening toys that had miraculously grown to greater size. The dwellings were sprinkled far into the distance across the landscape. Beyond them was the base of the mountain, magnificent and glorious in its crystal purity of white, and the striations, vertical and diagonal, that flashed brilliantly with black jet and peacock green.

Rawson knew them for mineral intrusions, and knew that the mountain was only one crystalline mass of all the quartz formation that made of the world's inner core a gigantic geode, gleaming in eternal brilliance under the glow of the central sun. And still, in it all, Dean Rawson had seen a lack without which perfection could not be complete.

"Where is Loah?" he asked of Gor. "I thought—I had hoped...."

Something in Gor's face told Rawson that his companion was troubled. "She refused to come," he said. "But the wish of one of the great ones from the Land of the Sun is a command." He shouted an order before Rawson could put in a protest. A man darted away.

"Always happy, my little Loah-San," said Gor. His eyes held a puzzled look. "Always until now. And now she weeps and will not say why. Come, we will walk more slowly. There were questions you wished to ask. I will answer them as we walk."

"Questions?" exclaimed Rawson. "A thousand of them."

And now for the first time since, at the top of a barren peak, in the dark of the desert night, his wild journey had begun, he found answers, definite and precise, to the puzzles he had been unable to solve.

Their speech—their language—how was it they could talk with him? He fired the questions out with furious eagerness, and Gor replied.

As to their speech—the Holy Mountain itself would explain. And yes, truly, this was the center of the world, or the sun above them was. The central sun did not attract, but instead repelled all matter from it—all things but one, the sun-stone, of which Gor would speak later.

Rawson pounced upon that and demanded corroboration.

"All the power of earth tends to draw every object to its center, yet we're here on an inner surface. We're walking actually head down. And our bodies, every stone, every particle of matter, ought by well-known laws to fall into that flaming center. But we don't! That proves your point—proves a counter gravitation. Then there must be a neutral zone. A place where this upward thrust is exactly equalled by gravity's downward pull.

"The zone of fire," said Gor. "You passed through it. Did you not see?"

"Saw it and felt it!" Rawson's mind leaped immediately to the next question.

"And we must have come through it at, surely, a thousand miles an hour. What drove us? That shell must have gone in from here. I can understand its falling one way, but not two. We should have come to rest in that very spot—and we'd have lasted about half a second if we had."

"Oro and Grah," said Gor. "Oro, the sun-stone, and Grah, the stone-that-loves-the-dark. But they are not stones, neither are they metal. We find them deep in the ground, clinging to the caves. A fine powder, both of them."

"Still I don't get it," said Rawson. "You drive that shell in from here, and then you drive it back again."

"That, too, I will explain later. It is simple; even the Dwellers in the Dark — those whom you call the mole-men — have Oro and Grah to serve them."

Gor launched into a long account of their tribal legends, of that time in the long ago when an angry sun god had driven his children inside the earth; of how Gor, and the son of Gor, and his son's sons tried always to return.

Rawson was listening only subconsciously. They were circling the white mountain, ascending its lower slope. Now he could see beyond it as far as the land extended, and he was startled to find this distance so short. They were on an island, ten miles or so in length, and beyond it was the sea; he must ask Gor about that.

"It is all that is left," said Gor, when Rawson interrupted his narrative. "Once the land was great and the sea small — this also in the long ago — but always it has risen. The air we breathe and the water in the sea come from the central sun. The air rushes out, as you know; the water has no place to retreat."

Again he took up his tale, but Rawson's eyes were following the upward curve of that sea. They, seemed to be in the bottom of a great bowl; he was trying to estimate, trying to gage distance.

"... and so, after many generations had lived and died, they found the Pathway to the Light," Gor was saying. "It is our name for the shaft through which you came. This was thousands of your years ago, when he who was then Gor, and the bravest of the tribe, descended. Even then they were workers in metal and they knew of Oro and Grah. They were our fathers, the first People of the Light."

Rawson had a question ready on his tongue, but Gor's words suggested another. "That shaft," he said, "the Pathway to the Light — do you mean it extends clear up to the mole-men's world? Why don't they come down?"

"To them the way is lost; the Pathway is closed above the zone of fire. That other Gor did that. And those who remained — the mole-men — have forgotten. They could break their way through if they knew — they are master-workers with fire—but for them the Pathway ends, and below is the great heat. But we know of a way around the closed place, the hidden way to the great Lake of Fire."

"They could break their way through if they knew!" repeated Rawson softly. For an instant he stood silent and unbreathing; he was remembering the ugly eyes in a priest's hideous face. The eyes were watching him as the White Ones took him away.

He forced his thoughts to come back to the earlier question. "What," he asked, "is the diameter, the distance across the inside world? How far is it from here to your sun? How many miles?"

"Miles?" questioned Gor. "We know the word, for the Mountain has told us, but the length of a mile we could not know. This I can say: there were wise men in the past when our own world was larger. They worked magic with little marks on paper. It is said that they knew that if one came here from our sun and kept on as far again through the solid rock, he would reach the outside—the land, of the true sun, from which our forefathers came."

Rawson nodded his head, while his eyes followed that sweeping green bowl of the sea. "Not far off," he said abstractedly. "Two thousand miles radius—and the earth itself not a solid ball, but a big globular shell two thousand miles thick. I could rig up a level, I suppose; work out an approximation of the curvature."

From the smooth winding path which they had followed there sounded behind them hurrying footsteps; a moment later Loah stood beside him.

Her eyes gave unmistakable corroboration of what Gor had said of that torrent of tears, but she looked at Dean bravely, while every show of emotion was erased from her face. "You sent for me," she said.

And Rawson, though now he knew he could speak to her and be understood, found himself at a loss for words.

"We wanted you with us, Gor and I," he began, then paused. She was so different from the girl whose smiling eyes had welcomed him. The change had come when he spoke those first words on his arrival, and now she was so coldly impersonal.

"I wanted to thank you. You saved my life; you were so brave, so...." Again he hesitated; he wanted to tell her how dear, how utterly lovely, she had seemed.

"It was nothing; it has pleased me to do it," she said quietly, then walked on ahead while the others followed. But Rawson knew that that slim body was tense with repressed emotion. He had not realized how he had looked forward to seeing again that welcoming light in her eyes. He was still puzzling over the change as they entered a natural cave in the mountainside.

A winding passage showed between sheer walls of snow white, where giant crystals had parted along their planes of cleavage. Then the passage grew dark, but he could see that ahead of them it opened to form a wider space. There were lights on the walls of the room, lights like the one that Loah had carried. And on the floor were rows of tables where men were busy at work, writing endlessly on long scrolls of parchment.

"The Wise Ones," Gor was saying. "Servants of the Holy Mountain." Yet even then men knelt at Rawson's coming as had the other more humble people. They then returned to their tables, and in that crystal mountain was only the sound of their scratching pens and the faint sigh of a breeze that blew in through a hidden passage to furnish ventilation.

Yet there were some at those tables whose pens did not move; they seemed to be waiting expectantly. One of them spoke. "The time is near," he said. "Are the Servants prepared?"

And the waiting ones answered: "We are prepared."

Rawson glanced sharply about. "What hocus-pocus is this?" he was asking himself. Still the silence persisted. He looked at the waiting men, motionless, their heads bent, their hands ready above the parchment scrolls. He saw again the white walls, the single broad band of some glittering metal that made a continuous black stripe around walls and ceiling and floor.

"What kind of ore is that?" he was asking himself silently. "It's metallic; it runs right through the mountain. I wonder—"

His idle thoughts were never finished. A ripping crash like the crackle of lightning in the vaulted room! Then a voice—the mountain itself was speaking—speaking in words whose familiar accent brought a sob into his throat.

"Station K-twenty-two-A," said the voice of the mountain, "the super-power station of the Radio-news Service at Los Angeles, California."

It's tuned in!" gasped Rawson. "Tuned in on the big L. A. station! A gigantic crystal detector! Those heavy laminations of imbedded metal furnish the inductance." Then his incoherent words ended — the mountain was speaking.

"Radiopress dispatch: The invasion of the mole-men has not been checked. Army Air Force fought a terrific engagement about midnight, last night, and met defeat. Over one hundred fighting planes were brought down in flames. Even the new battle-plane type, the latest dreadnoughts of the air, succumbed.

"Heavy loss of life, although civilian population of three towns had been evacuated before the mole-men destroyed them. Gordon Smith is reported killed. Smith was associated with Dean Rawson in the Tonah Basin where the mole-men first appeared. With Colonel Culver of the California National Guard, Smith was returning from Washington in an Army dreadnought which crashed back of the enemy's lines."

Rawson's tanned face had gone white; he knew the others were looking at him curiously, all but the men at the tables whose pens were flying furiously across the waiting scrolls. Before him the face of Loah, suddenly wide-eyed and troubled, swam dizzily. He could scarcely see it — he was seeing other sights of another world.

"They're out," he half whispered. "The red devils are out — and Smithy — Smithy's gone!"

CHAPTER XX
TALONED HANDS

Simple, pastoral folk, the People of the Light! In their inner world, a vanishing world, where nearly all of what once had been a vast country was now covered by the steadily encroaching sea, they had resisted the degeneration which might easily have followed the destruction of a complex civilization. Living simply, and clean of mind, they had clung to the culture of the past as it was taught them by their Wise Ones. And now the People of the Light had found a new god.

Not that Dean Rawson had asked for that exalted position; on the contrary he had tried his best to make them understand that he was only one of many millions, some better, some worse, but all of them merely humans.

His speaking the language of the holy mountain had convinced them first. But when old Rotan, oldest and grayest of the mountain's servants, went into a trance, then Rawson could no longer escape the honors being thrust upon him.

"The time of deliverance is at hand," old Rotan said when he awoke. His voice that so long had been cracked and feeble was suddenly strong, vibrant with belief in the visions that had come to him.

They were in the inner chamber of the white mountain, where Dean Rawson, heartsick, lonely and hopeless, had spent most of his time listening to the voice from the outer world. Gor was there, and Loah; and the writers had left their desks to gather around old Rotan, where now the old servant of the mountain stood erect, his glistening eyes fixed unwaveringly upon Rawson.

"Listen," he commanded. "Rotan speaks the truth. Never shall the People of the Light return to the outer world; it is here we stay. For now our world which is lost shall be returned to us." His eyes, unnaturally bright, met the wondering gaze of his own people gathered around, then came back to rest again upon Rawson.

Dean—Rah—Sun!" he said. "'Rah'—do you not see? It is our own word, Rah—the Messenger! Dean—Messenger of the Sun! The sun-god has sent him — he will set us free. He will restore our lost cities. The People of the Light will spread out to fill the new land; they will multiply, and once more will be a mighty nation, living happily as of old in their own lost world.

"Dean!" he called. "Dean — Messenger of the Sun!" He was drawn to his full frail height, his arms outstretched. But Rawson saw the old eyes close, sensed the first slackening of that tense body; it was he who sprang and caught the sagging figure in his arms, then lowered the lifeless body to the floor of crystal white.

Even happiness can kill. A feeble heart can cease to beat under the stress of emotions too beautiful to be borne. And Rotan, wisest of the wise, had passed on to serve his sun-god in another world.

And thereafter, Rawson, Dean-Rah-Sun, was undeniably a god. But he wondered, even then, while the others dropped to their knees in humble worship, why Loah, her eyes brimming over with tears, had broken suddenly into uncontrollable sobs and had rushed blindly, swiftly, from the room.

To Rawson the unwavering, simple faith of the White Ones was only an added misery. Rotan's vision was accepted by them unquestioningly; their adoring eyes followed Rawson wherever he went, while the children carpeted his path to the holy mountain with golden flowers.

And there Rawson would sit, cursing silently his own helplessness, while the voice of the mountain told of further devastation up above. His plans for leading a force against the mole-men were abandoned. On the island, all that was left of this inner world, were only some two thousand persons, men, women and children. And the children were few; the population had been rigorously kept down. Their present number was all that the island would support, though every possible foot of ground was tilled.

"Only a handful of them," Rawson admitted despondently, "and not a weapon of any sort. They've kept by themselves. Only Loah and a few of the others had enough curiosity and nerve to scout around where the mole-men live. She even understands their talk! Lord, what I'd give for a thousand like her, a thousand men with her nerve! Then, with weapons, and means of transportation...." But at that he stopped, aware of the futility of all such thoughts.

He had tried to talk to Gor, tried to tell him of his own limitations. And Gor had only smiled pleasantly and repeated "Rotan has spoken. It will come to pass!"

Ceaselessly his thoughts revolved about the hopelessness of his situation. He was alone. Whatever was to be done he must do single-handed—and there was nothing he could do! But he would not admit to himself that the aching loneliness came to a focus in the memory of a girl's smiling eyes, the touch of her soft hand.

"They're fighting up there," he argued, "fighting for their lives, and I can't help. What right have I to think of Loah or myself?" In spite of which he sprang abruptly to his feet, left the mountain and the voice of the mountain behind him, and went in search of the girl.

"I've got to make her understand," he exclaimed. "I've got to have someone to talk to. But I can't make her out. She's so confoundedly respectful—acts as if I were a little tin god. And yet—she wasn't always that way!"

At the home of Gor he found Loah, slim and beautiful as always. She had just come from the bath. The creamy texture of her skin had flushed to rosiness in the cold fountain. Her jeweled breast-plates sparkled. A cloth that shone like silk enwrapped her hips in soft folds of pale rose and hung in an absurd little skirt. She might have been the spirit of youth itself, a vision of loveliness; yet Rawson felt an almost uncontrollable desire to take her in his two hands and shake her when she bowed humbly and treated his request as if it were a royal command.

"To walk with Dean-Rah-Sun! But certainly, if that is his wish!"

In silence they left the village and walked toward the island's end where Rawson had emerged from the under-world.

The island was not large. On either side were low hills, mere knolls, of white crystal, where, in every hollow, men and women were harvesting strange grain. Between the two ranges of hills were flat fields of green, reaching out toward the point some three miles distant.

Rawson made no attempt to talk as he led Loah along the roadway that cleft the green expanse in half. Other workers were there, and Dean acknowledged their smiling, worshipful salutations. He did not want to talk now; he wanted to find some place where he and Loah could be by themselves. There was so much he must tell her. He must try to make her understand. And after that, perhaps, with her help, he could find some way to be of aid to his own beleaguered people—something he could do even single-handed.

Where the fields ended, and from there on toward the point, had been an expanse of glistening white. Rawson remembered it plainly. So now, when he found it a place of flaming crimson, he stared in amazement. Across the full width of the valley a brilliant carpet had spread itself, a covering of flowers. A blossoming vine had sprung up in the few days since his arrival and had woven a thick mat of vegetation.

He wanted to go on out to the extreme end of the point. There they would be alone. But Loah objected when he started to enter the red expanse.

"No!" she said in quick alarm. "We must not cross. It is the Place of Death. We will go around it, following the hills."

"We crossed it the other day when it was a plain of white salt," argued Rawson.

"But now the flowers have come. Even now it might be safe—but when they die then nothing can cross here and live."

Loah could not give the reason. Dean gathered from what she could tell that a gas of some sort was formed, perhaps by the decomposing vegetation. Perhaps it combined with the sparkling white shale. But all this was of no consequence compared with his own problems. He did not argue the matter but followed where Loah led.

"Where is the shell?" he asked, when they stood at last near the open mouth of the great shaft into which the air was rushing. "Where is the machine that we came here in? I wanted to see it—thought perhaps I could use it later on.

"The jana—the shell, as you call it—is safely locked in a great room of Gor's house. Not all understand its use; it must be kept away from careless hands."

Then Rawson put that thought aside. He took Loah's hand and led her some distance away toward the shore. Beyond a rocky, crystalline mass, where fragments had been heaped, the sound of the rushing air was lost; only the flashing emerald waves whispered softly on the shore beyond. And there in that quiet place, under the brilliance of the central sun, Rawson told her of himself and of the great outer world. He told her of his work, of everything that had happened, of how he was only one of many millions of men and women like, and yet unlike, the People of the Light. And at last he knew that she understood.

He had spoken softly, though he knew there were no other listening ears. Loah had been seated before him on one of the white blocks. She rose to her feet. Her eyes were troubled. Vaguely he sensed behind them a conflict of emotions.

"I must think," she said. "I will walk by myself for a time; then I will return."

Rawson reached for her hand. "You're a good sport," he said huskily. Then he felt the trembling of that hand in his; and, as if it had been an electric current, his own body responded.

Shaken in every nerve, his poise deserted him. He could not think clearly. He knew only that that horrible loneliness was somehow gone. By force of will alone he kept his arms from reaching out toward that radiant figure. Instead, he raised her hand toward his lips.

She withdrew it sharply. "No," she said, "our Wise Ones were mistaken. For years they have listened to the mountain; they have written down its words. Slowly they have learned their meaning. A kiss, they said, was a symbol of love in your world. They were mistaken—as was I. Now I will walk alone for a time."

Rawson let her go. She seemed hardly looking where she went; her eyes were downcast. She moved slowly around the sheltering rock and on toward the level ground and the rushing winds of the shaft.

His own thoughts were in a whirl, too confused with emotion for clear thinking. "A symbol of love!" And back there in that cave world she had pressed her lips to his hand. Then they had come here, and he had been transformed to a god, a being who could never have more than an impersonal affection for one as humble as she.

The rising flood of happiness within him was abruptly frozen, changed to something which filled his veins with ice. For, from beyond the crystal barrier that hid Loah from his view, her voice had come in one single cry of terror. Then, "Dean!" she called. "Dean San!" But by then, Rawson was throwing himself madly around the barricade of rocks.

Like a sensitized plate when the camera's shutter is opened a merest fraction of a second, Rawson's brain took the imprint of every detail that was there. The black mouth of the shaft, and, on the rock beside it, something metallic, brilliantly gleaming—a flame-thrower! Beyond the pit was Loah, half crouching, her slim body tense as if checked in mid-flight. She had been running toward him, coming to warn him. And between her and the shaft, his back turned squarely toward Rawson, was the hideous figure of a mole-man, one of the Reds! His grotesque, pointed head was bent forward toward the girl; his arms were reaching, the long fingers like talons.

Rawson did not know when he called the girl's name. But he knew the instant that he had done it and he knew it was a mistake. He should have crept quietly, seized the weapon—and now his feet tore madly on the white rock floor as he raced toward the shining implement of death. From beyond, the red figure, whirling at his call, leaped wildly for the same prize.

The taloned hands were on the flame-thrower first. Rawson saw the red body straighten, saw the weapon swing, glistening in air, swinging over and down. From its tip green fire made a straight line of light.

He leaped in under the descending flame, felt the nozzle of the projector as it crashed upon his right shoulder and the green fire spat harmlessly beyond his back. That last spring had thrown him bodily against the red monster. They were both knocked off balance for a moment, then Rawson caught himself and swung with his left. He set himself in that fraction of a second, felt the first movement of that shining, crook-necked tube that meant the green flame was being drawn back where it could reach him; then his fist crashed into a yielding jaw.

Not five feet from the brink of that nearly bottomless shaft he stood wavering in the rush of air. He knew that the ugly red figure had toppled sideways, that the weapon had fallen with him, the blast swinging upward in a vertical, hissing arc — then man and weapon had dropped silently into the pit.

He was alone, save for the girl, who, her eyes wide with horror, threw herself upon him and clung trembling, while she murmured incomprehensible endearments in her own tongue wherein his own name was mingled: "Dean, dear! My own Dean-San!"

But the mole-men! Dean Rawson's mind was aghast with the horror of it: the mole-men had now found the way.

CHAPTER XXI
SUICIDE?

Gordon Smith, sometimes known as Smithy, was to remember little of the happenings that followed the crash of the big Army dreadnought. It was Colonel Culver who dragged him from the pilot-room wreckage, Colonel Culver and one of the pilots whom he had restored to consciousness. They lowered Smithy carefully to the ground, then explored the rest of the ship.

Their hands were red when they returned — and empty. Captain Farrell and the rest of the crew had ceased to be units of the United States Army Air Force; henceforth they would be only names on a casualty list grown ominously long.

"Stood plumb on her tail," said the pilot, staring at the wreck. "They hit us just once, and the left wing crumpled like cardboard. Last I remember was pulling her up off the trees." He stared at the mass of twisted metal and the center section where the wing had torn loose; it stood upright, almost vertical, resting on the crushed tail.

"Funny," said the pilot in the same flat, level tone that seemed the only voice he had since that last pull on a whipping wheel. "Damn funny — mostly we get it first up there."

"Come here!" snapped Colonel Culver. "Lend a hand here with Smith; we've got to carry him. And don't talk so loud — those red devils will be out here any minute."

Smithy was taking a more active interest in his surroundings when he sat a week later in the Governor's office.

"There's a detachment moving in there from the south," said the Governor. "We're going to follow your advice, to some extent at least. We're sending troops to Tonah Basin. If the top of that dead crater is closed they will blast it open; then a scouting party's going down. Call it a reconnaissance, call it suicide — one name's just as good as the other. Colonel Culver, here, is going. But you know the lay of the land there; you could be of great help. How about it?"

"Are you asking me?" Smithy inquired.

He stood up, flexed his arms, while he grinned at Colonel Culver. "Hinges all greased and working! As a flier, Colonel, you're a darn good first-aid man. I'll say that! When do we start?"

Which explains why Smithy, some time later, hidden under the grotesque disguise of a gas mask, was one of fifty, similarly attired, who stood waiting about the black open maw in the great cinder-floored crater of one of the peaks that surrounded Tonah Basin.

Night. And the big stars that hang so low in the black desert sky should have been brilliant. They were lost now in the white glare that streamed upward. The crater was a fortress. Around the circle of the entire rim, on the inner side of the rough crags, men of the 49th Field Artillery stood by their guns. Lookouts trailed their telephone wire to the higher peaks, where they perched as shapeless as huddled owls; and, like owls, their eyes swept the mountain's slopes and the desert at its base, where the searchlight crews played long fingers of light incessantly — and where nothing moved.

But the empty silence of the desert was misleading, as the men in the crater knew.

They had begun arriving with the earliest light of morning. Smithy had come in with the first lot. And when the first big auto-gyro transport had settled and risen again from the crater, another had taken its place, and another and many others after that.

That first crew had been a machine-gun battalion, and Smithy had smiled with grim satisfaction at the unhurried way in which their young captain had snapped them into position without the loss of a second. And their guns, Smithy noticed, were trained inward upon the crater itself.

Inside that protecting circle the other transports landed one by one: men, mobile artillery, ammunition cases, big searchlights, and a dozen engine-generator outfits. The last transports brought in strange cargo — short sections of aluminum struts with bolts and splice plates to join them together: blocks, and tackle and sheaves; then spools of steel alloy cable at least ten miles in length.

From the last ship they took a hoisting engine and an assortment of aluminum plates and bars which were bolted together by waiting mechanics, and which grew magically to a crude but exceedingly substantial elevator, on which fifty men, by considerable crowding, could stand.

Only a floor of bolted plates, with corner posts and diagonal bracing and a single guard rail running around the four sides — but for the first time

Smithy began to feel that he was actually going down; that this was not all make-believe, or a futile gesture. He would stand on that platform; he would go down where Dean had gone. And then.... But what would come after he knew he could never imagine.

A little crane swung the first metal work into position above the shaft. One end of the assembled framework of aluminum alloy dragged loosely on the ground; the other end swung out and projected above the shaft, swayed for an instant—and then came the first direct knowledge of the enemy's presence. The end of a metal strut, though nothing visible was touching it, grew suddenly white hot, sagged, then broke into a shower of molten, dazzling drops that rained down into the pit.

"Good," said Colonel Culver, who was standing beside Smithy. "Now we know they are there—but it means we will have to go down there with our gas masks on."

To Smithy it was not immediately apparent how gas masks were to protect them from the deadly invisible ray. He got the connection of thoughts when a bomb was slid over the edge. The dull thud of the explosion quickly came back to them.

"They popped that one off in the air—hit it with their heat ray," said a cheerful voice beside them. "But the phosgene will keep on going down. Give them another!"

The interval this time was longer. "Now for a dirty crack," said the cheerful voice. "Time this one."

A youngster nearby snapped a stop-watch as the bomb was released. He held some printed tables in his hands. Odd receivers from which no wire led were clamped over his ears. This time the dull thud was long in coming. It was hardly perceptible when the young man with the stop watch announced: "Fifty thousand feet, sir."

"Give 'em another. Time it again." A second high explosive bomb was released.

"Fifty thousand feet, sir."

"Good. That measures it. And those last bombs have knocked the devil out of whatever machinery they've got down there. Now we'll give them a real taste of gas. Two of the green ones there, men. Put ten miles of cable on the drums. Get that hoisting frame into place."

But night had come, though searchlights outside the crater and floodlights within had robbed the night of its terror, when Smithy, with

Culver beside him, climbed over the guard rail of the lift that hung waiting just over the pit.

A gas mask covered his entire face. Through its round eye plates he looked at the others who crowded about him. Grotesque, almost ludicrous—twenty men, armed with clumsy sub-machine guns; the others would follow later. A searchlight was on a tripod at the center, and a spool of electric cable.

The light sizzled into life and swung slowly about. Then the platform jarred, and the spool of cable began slowly to unwind. Beside him Colonel Culver was returning the salute of an officer outside on the ashy ground. Smithy raised his hand, but the brink of that pit had moved swiftly up—there was nothing before him but a glassy wall.

Reconnaissance? Suicide? One word was as good as another. But he was going down—down where Dean Rawson had gone—down where there was a debt to be paid.

CHAPTER XXII
THE RED-FLOWERING VINE

Rotan," said Gor slowly, sadly, "was wrong. His vision was not the truth. The Red Ones have come. And now—we die."

"Without a fight?" Rawson demanded incredulously.

"We are not a fighting people. We have no weapons. We can only die."

Rawson turned to Loah. They were inside the mountain, and the servants of the mountain, with terror and dismay written plainly on their faces, were gathered about. "At the Lake of Fire," said Rawson, "when you saved me, there was an explosion and clouds of white fumes. What was it?"

"It was like water," Loah said. "We found it deep inside the earth in a place where it is very cold. When warmed it turns to white clouds. We threw a flask of it on the hot rocks, hoping to reach you while they could not see."—she paused and shook her head slowly—"but we can get no more. The Pathway of Light is closed to us, now that the Red Ones are there."

"Liquefied gas of some sort," said Rawson briefly, "caught in enormous rock pressure. But that's out! Now what about this Place of Death? There's an idea there."

The White Ones were numbed with fear, but Loah and Gor accompanied him when Rawson returned to the red field. The flowers were still in bloom; they waved gently in the breeze that blew always from the mountain across the fields and out toward the point, where even now dark figures could be seen near the mouth of the shaft.

"It will be many of your days," said Loah, "before the flowers die. If you thought to trap the Red Ones in the Place of Death, there will not be time...." But Rawson had left them; he had advanced into the scarlet field and dropped to his knees.

He was crushing the vines in his hands, grinding them into the white, salty earth underneath. Then he passed his hands guardedly before his face as if to detect an odor.

Loah and Gor saw him shake his head slowly while he spoke aloud words that they could not understand. "Cyanide," Dean Rawson was saying. "It's a cyanide of some sort — releases hydrocyanic acid gas. I could have rigged a generator, though I've forgotten about all of my chemistry — and now there isn't time." Off in the distance the dark figures still moved near the end of the point.

He made no effort to conceal his dejection as he returned. The edge of the Place of Death made a winding line across the scant half mile of valley where the green fields ended abruptly.

Dean stepped high over the stone trough a half mile long that marked that dividing line. There was water in it; it was part of their irrigation system. A little beyond, in the midst of the green, stood a tiny flat-topped knoll on which he knew was a pool that supplied the crude system. Beyond it Loah and Gor were waiting.

Gor read the look on Rawson's face. "It is useless," Gor said. "And now I have decided. The People of the Light must die — but not in the fires of the Reds. With my people I shall walk into the sea."

And Rawson could not protest. He could only follow as Gor turned back toward the village and the mountain beyond.

From a spur on the mountainside Rawson could see the full length of the island. One way lay the village; beyond it the green fields; then the wide scarlet band of the Place of Death. And beyond that the little crystal hills and the valley between that led out to the point. It was now dark with massed clusters of bodies, red even at that distance. He could even see the glint of metal from time to time.

And behind the mountain were the People of Light, where Gor was only waiting for the attack to lead them out to the island's farther end and then on to a kindlier death in the emerald sea. Only Loah was with Dean, although there were others of the White Ones not far away, watching, ready to warn Gor when the attack began.

Not an hour before, Rawson had stood in the inner chamber and had listened to the mountain as it repeated the words of a far-distant man: "Attack of the mole-men growing increasingly ferocious ... heat-ray projectors — almost invincible ... our forces have entered the Tonah Basin — they are descending into the crater. But whether warfare can be carried on advantageously under ground is problematical...." Rawson unconsciously gritted his teeth behind his set lips as he watched the Reds.

He knew why they had been so slow in attacking. They must have a carrier of some sort, a shell like that of Loah's, and they were bringing their fighters one shell-load at a time. When the entire force was ready they would attack. And Rawson was convinced that this force would be limited in number.

"They'll have plenty to keep them busy up there," he argued. "If only we could wipe out this one lot we could prepare to defend ourselves." And now, standing on the side of the mountain, he startled Loah with the fury of his sudden ejaculation.

"Fool! Quitter! Waiting here for them to come and get you! There's one chance in a million—" Then he was rushing at full speed along the roadway that circled the mountain toward Gor and the terrified throng.

The waiting savages must have laughed, if indeed laughter was possible for such a race, at sight of the White Ones creeping timidly down. Off a mile and more they could see them harvesting their strange crop—harvesting!—storing up supplies of food, no doubt, when the mole-men with their flame-throwers would reap the harvest so soon!

But in a crimson field Dean and Gor and Loah led the others where they swarmed across the Place of Death, gathering huge armfuls of the red-flowering vine, carrying them to the village and returning for more. Where they trod it was as if peach pits were crushed beneath their feet. And there was a curious fragrance which Rawson told them not to breathe, but to keep their faces always into the wind.

Their hands and bodies were sore and burned by the strong juice of the vines. They stopped often to cast apprehensive glances at the distant group of red figures, and always Rawson drove them in a frenzy of haste. At last he made them move the long trough of stone beyond the edge of the green field and over into the Place of Death.

Rawson kept no track of the time. The voice of the mountain was his only measure of hours in a world of perpetual day. But more hours—another day, perhaps—had passed when the Red force at last began to move.

They did not spread out wide across the valley, but formed a straggling line that was denser toward the center. They could not know what opposition they would meet; for the present they would stay together. Above them as they came were twinkling lights of pale-green fire.

The radio had spoken of heat rays; Rawson wondered if that meant some newer and more horrible instrument. But he saw nothing but the flame-throwers in the armament of this force.

He was waiting by the irrigation pool, hidden for the moment behind the little knoll. Loah was with him; he had tried in vain to induce her to stay with Gor and the others who were waiting beyond the mountain.

There were watchers, some of them within hearing, whose voices relayed the news of the enemy's advance. Then they ran; panic was upon them.

"*Tur – gona!*" they cried, "*Nu – tur – gona!* We die! Quickly we die!" Rawson heard the shout carried on toward the hidden throng.

Cautiously he peered from the little knoll. They were coming. Already they were trampling the remaining red blooms on the farther edge of the field. But he waited till they were halfway across before he leaped to the top of the knoll, grasped a pole he had placed there in readiness and rammed it down through the pool, turbid yellow with the juice from the vines, and broke open the outlet he had plugged in the base.

One green light slashed above his head. One flicked at the knoll near his feet, where green growing things burst into flame – then he threw himself backward down the short rocky slope while the stones tore at his nearly nude body. He sprang to his feet and held Loah close. On either side of the knoll was a holocaust of flame where green lights played. He waited breathlessly. The fires brought in a little back draft of air, the scent of peach pits was strong – and then the green lights ceased. The unripe grain of the fields smoldered slowly.

Then Rawson stepped from his hiding and stared out at the Place of Death.

Nearby was a huddle of bodies. On either side, in a long, straggling line, they lay now on the ground – a windrow where Death had reaped. The flames of their weapons still in action were all that moved. The white earth turned molten wherever those flames struck.

Farther off there were red things that were running. The yellow liquid from the pool, charged with the acid of the vines, had been slow in flowing out through that long trough. The savages could only see that their fellows had fallen. Some mystery, something invisible and beyond their comprehension had struck them. They ran toward the center at first, then turned and fled – and by then the soft air blowing gently about them had brought that strange fragrance of death. Then they, too, lay still.

From the distance came faintly a booming chant, two thousand voices raised in unison. "*Tur – gona! Nu – tur – gona!*" The last of a once mighty people were marching to their death.

Rawson and Loah turned with one accord. Victory was theirs, but there was no time to taste the fruits of victory. They ran with straining muscles and gasping breath toward the distant mountain and the marching host beyond.

My plans are made," Rawson spoke quietly. "I must go. I shall take the shell – the jana – and go back to the mole-men's world. I shall go alone, and I shall die, but what of that?" His eyes lit up for a moment. "I'll try to find *Phee-e-al* first. If I can get him before they get me, that will help."

They were standing on the mountain's lower slope, Gor and Leah and the servants of the mountain gathered near. Below, the White Ones were massed in worshiping silence. Had not Dean-Rah-Sun saved them? And now what else would come to pass?

The same question had been asked by the Wise Ones, and now Rawson turned and spoke to them. "Rotan was right," he told them. "His vision was true. There is work I must do here before I go. Your lands, or some of them at least, will be restored. And you will be safe forever from what we have seen to-day. Gor will lead you wisely, and Loah...." His voice faltered; he had kept his eyes resolutely away from the slim figure of the girl, who had been wordless, scarcely breathing. Now she stepped swiftly before him.

"You must go, Dean-San," she said gently. He knew it was a term of endearment. "You must go if you say you must. But you do not go alone, nor die alone. Long ago the voice of the mountain spoke beautiful words. I know now it was one of your priests telling of a woman of your own race. Always have I remembered. 'Wheresoever thou goest, I shall go; thy people....'"

But Dean Rawson had gathered the slender figure, starry-eyed and sobbing into his arms.

CHAPTER XXIII
ORO AND GRAH

"The Place of Death!" said Dean Rawson. "Whoever named it had the right idea."

As part of their titanic plan, Rawson and Loah-San return to sacrifice themselves in the flaming caverns of the Red Ones.

He looked out across the wide stretch of ground with its covering of white salt almost entirely stripped of the carpet of vines. The bodies of the mole-men lay where they had fallen; their flame-throwers still tore futilely at the earth or stabbed upward in vain, thrusting toward the green-gold sun that shone pitilessly down.

"Still I do not understand," said Gor. "My people pressed the strong, burning water from the vines and poured it into the pool as you directed. But the Red Ones did not touch it—how could it burn them?"

"I'll say it was strong!" said Rawson. He looked at his hands, red and burned where the liquid had touched. "And it got stronger by standing. It was an acid, and when it touched the white earth a gas was formed—hydrocyanic acid gas. And that's nothing to fool with."

He walked cautiously out where the liquid had been poured over the white ground. No odor remained; the air was clean. Then he picked up one of the flame-throwers and experimented with it until he found the sliding sleeve that shut off the blast.

"All right," he called to Gor. "Bring on your men; we've got to clean up this place and get rid of the bodies before the sun gets in its work. They're the ones that will go into the ocean instead of you." He moved carefully along the straggling line of bodies, salvaging the weapons and turning off their fearful blasts.

They worked and slept and worked again before their gruesome task was done and Rawson was ready to begin the other work that he had in mind.

Beside the mouth of the great shaft, resting on the rocks, was a cylinder, almost exactly a counterpart of the one Loah had used. But this was larger—fully fifty of the red savages could have crowded inside.

"It is the only one they had," said Loah. "I have seen, and I know."

"But they can make more," Gor argued. "This one and the one we have," he told Rawson, "were made thousands of years ago. There were masters of metal-work among them, and they had learned to use Oro and Grah. Even then the people were divided. He who was then Gor and his followers fought with the others. But he left them one *jana*—this very one here. Then Gor followed the Pathway to the Light, though he sealed it as you know. But—but they will build others. Sooner or later they will come."

"I think not," said Rawson. "Now what about this Oro and Grah material? What was it you called them—the Sun-stone and the Stone-that-loves-the-dark? I must know how they work." But Loah was reluctant to experiment with the *jana* of the Reds; she had her own shell brought instead—and then Rawson learned the secret of what seemed its miraculous flight.

A cylindrical metal bubble, just buoyant enough to lift itself above the ground—Gor and some of the others brought it from the village. Gor brought, too, a little box which he carried with great difficulty.

"It is Grah," he said, when he showed Rawson a little scattering of black dust within the box. "Always it tries to fall back under the ground. Both Oro and Grah grow deep down near the Zone of the Fires; we find them in the caves, Oro on one side and Grah on the other. Oro is as heavy in its upward falling as Grah is in its downward.

"Then"—he pointed to the central vertical tube in the shell—"we put both of them in here, bringing it a few grains at a time. One falls to one end and the other to the other. And then, with these simple valves, we let out a little of whichever we wish—release it a grain at a time, if that is best. We let out a few grains of Grah, and Oro, being stronger, draws us upward; or we let a little of the Oro escape, and we fall downward swiftly. You see it is simple, as I said."

Rawson's reply was not an answer to Gor so much as it was an argument with himself. "Heavy," he said. "Specific gravity beyond anything we've ever known. Osmium, the heaviest substance we have, would be light as a feather compared to this. But wait. This Grah, as you call it, falls downward, but that means it falls toward the outside of the earth. With us it would be light—light! And Oro would be heavy. New substance—new matter! One feels only the attraction of our normal gravitation; the other doesn't react to that at all, but is driven outward with tremendous force by counter-gravitation, the repulsion of this Central Sun. You've used it cleverly, but we'd have done more with it up on top."

He was lost in thought for some minutes, muttering figures and calculations half aloud. "Two thousand miles from the Central Sun to us; two thousand more through the solid earth. And if that repelling force follows Newtonian laws it will decrease as the square.... But, coming down from up on top, normal gravity would decrease directly as the distance!" He made scratches with one small stone upon a larger one in lieu of paper and pencil, but, to his listeners, his muttered words could have meant nothing.

"Around six seventy-six hundred and seventy miles to the neutral zone, the Zone of Fire. And a column of water—it would carry on by, plug the shaft, check the back-pressure, and then...." For the first time since that night when the mole-men had poured out into the crater, his eyes were alight with hope, though his face seemed tense and grim. Then the lines about his lips relaxed; he smiled at Loah.

"I would like to investigate this under-world," he said, "—not very far down. Will you take me?"

The girl's adventurous spirit had led her on many exploring trips in that subterranean world. She laughed happily when Rawson told her what he wanted. "But, yes," she said; "of course I know such a place." And from some two or three miles below, after anchoring the *jana* securely, she led him through a winding tunnel where he knew he was steadily climbing.

It was a wide corridor that they followed, where the walls came together high above their heads; he could hardly see where they met by the light of Loah's torch. Now and then there were lateral passages, but they were narrow, hardly more than cracks; and Rawson, looking into them, nodded his head with satisfaction.

Occasionally his footsteps rang hollowly on the stone, and he knew that the floor was thin between this and other caverns below. "What an old honeycomb it is!" he exclaimed. "And we had it all figured as being solid. The weight is all here, of course, but it's concentrated in that red stuff down near the neutral zone. But anyway, Loah has shown me just what I wanted."

He had gathered a handful of little fragments, and, keeping count of his steps, had shifted a bit of rock to his left hand for every hundred paces. By this he knew they must have gone five or six miles when he reached the tunnel's high point. Many times it had widened. Here, too, was a cave more than a hundred feet across.

From the farther side the tunnel continued, pitching sharply downward, but Rawson did not explore farther. "I can seal that off with a flame-thrower," he said. "I've seen how they use them." Then he took Loah's light and looked with every evidence of approval at the rocky walls and the roof that seemed heavy with dew.

He had wondered about the air, but he found that it seeped through from that central shaft, although Loah told him that in some deeper passages the air was bad. Here, although it was moving gently, it seemed wet as if charged with moisture. Rawson, staring upward, felt a drop strike him in the face, dripping from the rocks above.

"It's a gamble," he said, "just a gamble. But the stakes are worth while. And now, Loah-San, we will return."

He made crude work with the flame-throwers at first but finally he got the knack, and the mouth of the tunnel beyond the big room was sealed. Then, with the help of Loah and some few of the others, he brought in more and more weapons of the Reds. He was curious as to their construction, but his curiosity had to go unsatisfied. They were only cylinders, so far as he could see, cylinders a foot long and six inches through, of some metal with the dull lustre of aluminum. But they were sealed, and he dared not cut one open with another flame-thrower for fear of what might come forth.

On the top of each cylinder a tube was connected that ended in a lava tip; but at the base of the tube, where it joined the cylinder, was a sliding sleeve that checked the flame to nothing when it was moved, or opened it to the full blast.

He had a hundred of them in the room when at last he was through — one hundred fearful instruments of destruction. And still he told no one of his plans; he only told Gor what he wanted done later on. "It may not work," he had to admit to himself. "I'm just guessing at the thickness of the rock and the power of these machines. It's a gamble, nothing but a gamble."

He arranged the flame-throwers in a circle along the outer wall. The tops of the cylinders were curved, but the bottoms were flat and they set solidly on the rock. But he tipped them backward and braced them firmly with fragments of stone until every crooked-neck tube was pointed upward and toward the center. Finally he was done.

It was only a matter of a few hours later when Rawson stood on the island's end by the mouth of the shaft. In his ears was the ceaseless rush of the air as it entered the pit; it was the only sound in a silent world. And for the first time there came overwhelmingly upon him a realization of what this moment meant.

The time had come. Loah was beside him, her lovely eyes unnaturally bright in her face from which all the blood seemed to have flowed. He felt the slight trembling of her body as she pressed against him; he knew she was struggling to keep back the tears. Then Rawson half turned with one final entreaty that she let him go alone; but he left the words unsaid — he had argued it several times before.

Before them stood Gor, then the Wise Ones, the Servants of the Mountain, deserting their post for the first time since the Mountain had been given a voice. Beyond them all the people of this little world were gathered.

It had seemed only a fanciful dream, this thought of going; in fact, he had been too busy, too pressed with his own preparations, to give it thought. Now he was learning to his own surprise how closely he had identified himself with this world and its people. It had given him Loah; it had been a haven, a sanctuary.

He let his eyes slowly take in the full splendor of that emerald sea, the shining land under a green-gold sun, the Mountain in white, crystal purity against a green-blue sky. And he was leaving it, he and Loah; they were going to—death!

"You will remember," he said to Gor. His voice sounded dull and heavy; it hardly seemed himself who was speaking. "You know the day and the hour. This is the nineteenth. It is now noon—twelve o'clock in my world. When the Voice of the Mountain says that noon again has come you will do as I said."

"The Mountain speaks without ceasing now," said Gor, "telling always of what the Red Ones do. We will count the hours as they pass. In twenty-four of those hours Gor will descend in the *jana* of the Reds to do as Dean Rah-Sun has commanded."

Rawson held out his hand. He was suddenly wordless. Then Loah threw herself into Gor's arms in one last passionate embrace—but it was she who entered the *jana* first.

"Come," she said to Dean. "Oh, come quickly, Dean-San!" Then he, too, stepped inside and made the heavy door fast.

Men of the White Ones had been holding the big cylinder down. But Rawson, staring through the window, saw that it was Gor's own hands that swung them out at last above the pit.

Their craft hung quivering for an instant in the rushing air; then Loah moved one of the levers a trifle and the blackness took them, and only the little bull's-eyes in the metal ceiling showed the fading glow of the Inner World, the home of the People of the Light, which their eyes never again would see.

CHAPTER XXIV
THE Bargain

Rawson had taken one flame-thrower with him. He tied it securely inside the shell so it could not shift with the changing gravity, or be accidentally turned on. Again he clung to the curved bar against the wall. Loah stood at the center, directing the craft.

Once again he floated in air, then found himself standing on what had been the ceiling of the room. The girl had released a considerable quantity of the lifting element in the *jana's* end, and now the black powder in the other end of the central tube was dragging them at terrific speed as it rushed away from the earth's center.

Over six hundred miles, Rawson had figured, from that inner surface to the neutral zone where the red substance of the earth, that was neither rock nor metal, under terrific pressures, glowed with fervent heat or formed pools like the Lake of Fire.

Perhaps a hundred miles thick, that zone of incessant energy, and their little craft tore through it at tremendous speed. Even so, he was gasping for breath in the heated room when the glow faded and again he swung over and down upon the floor as Loah checked the speed of the flying projectile and the little ship crept slowly up into the room where first he had seen it.

The first that he noticed was the absence of the roar. The *jana* drifted slowly to one side, and Loah let it come to rest upon the floor. Staring from the open door, Rawson saw the same familiar red walls and floor and the black opening of the shaft from which they had come. But the reverberating roar of the great organ-pipe was gone. He knew that the air, for the greater part, was driving on past through the upper shaft that was now open. The way was clear for them to ascend. He turned to the girl.

"If my figures are right, it's some thirteen hundred miles from here on. How did you get up there before?"

Loah pointed to the passage where the *jana*, on that other excursion, had been hidden. "We went through there," she said, "taking the *jana* with us. We went up many miles through a great crack, but it was not straight;

we had to go carefully till another passage opened through to the shaft far above where it was sealed."

"And the mole-men never found it?"

"Oh, yes," said Loah, "they must have known of the crack, but they did not know where it led. Its air was bad — a gas that choked; one could not breathe it and live. But in our little *jana* we were safe. They could not use theirs; it was too large. Besides, only the priests came down. They had their Lake of Fire, where they did horrible things. They did not know that the shaft began again below."

"O. K.," said Rawson, and closed the door.

"But I wish to get out," Loah protested, "to gather more of the Oro. We may need more, should we return."

"We will never need it," Rawson spoke softly. "From the time we left Gor we had just twenty-four hours to live. We must go on, and go fast."

They had no way of measuring time, and Rawson could only guess at the hours that passed while their little ship tore swiftly upward through the dark. He wondered if the occasional shrill shriek that followed the touching of their metal guides on the glassy walls could be heard up above.

Then, at last, Loah was driving the *jana* slowly while she held her light so it would shine through a window. Rawson had to restrain himself to keep from pacing the little room like a caged animal while the precious minutes slipped by. Now that the enemy was near he wanted nothing but to drive on up to the end of the shaft, come out into that world wherever the shaft ended, then try to fight his way through to the great hall where he hoped to find Phee-e-al. And his haste made him overestimate the passing time; their journey had been swifter than he knew.

"I may have passed it," Loah was saying doubtfully. "I may have come too far." Then she interrupted herself and sprang to the controls.

They drifted slowly back. "It is different now," Loah said; "the air rises more swiftly than before." She stared from the windows while she drove the *jana* slowly up and down, trying to bring it to equilibrium in the strong up-draft.

The air entered the shell through a little opening with the same pungent tang Rawson had noticed before. He had wondered about the air. Down near the neutral zone it was dense, yet he had not minded the pressure too greatly — and that had been puzzling.

"Rock pressure and air pressure," he had reasoned; "they are two different things. If the rock flowed, any air that it trapped would be squeezed to a liquid. But it doesn't flow — that red stuff is solid; so the air pressure is only the weight of the air column itself. But even that should be enormous."

He could only conclude that the lessened pressure came from that strange counter-gravitation, the repelling force from the center of the earth. Perhaps it tended to dissipate the molecules, held them farther apart, prevented their squeezing in together, and battering with a thousand little impacts on a point where one had hit before.

Their *jana* swayed gently as if the smooth air currents were disturbed and were drifting them sideways; and then, at last, Loah, peering from a window, sprang back and moved a lever. Beneath them was the softly-cushioned thud of the shell seating itself on firm rock.

They were in another of the interminable caves, Rawson found when he opened the door. The *jana* was resting a few feet in from the edge of the shaft. Cautiously they got out, but even without their weight it had a slight negative buoyancy.

"Oro is pulling more strongly than Grah," Dean said, and smiled. Already the names seemed familiar to him.

The two lifted the *jana* and carried it back some twenty feet more before Rawson realized how unnecessary this was.

"We'll never be using it again," he said. "If I've guessed right it will stay here as long as the rocks; if not — but we'll never know the difference anyway."

He took the flame-thrower from the car in sudden haste. "Quick, dear," he told Loah. "God knows when the end will come. Quick, show me the way."

Loah knew every step of the route that took them on and upward through a maze of twisting passages, and Rawson marveled at her sense of direction. She flashed her light at times — the little bar of metal that had in one hollow end a substance which absorbed the light-energy of the Central Sun. Rawson knew how it worked. Even the lights in the mountain room were taken out from time to time and exposed to the sunlight that brought them back into glowing life. He had seen similar phenomena on earth. But, for the most part, Loah kept the little metal cap in place on the end of her torch, and they moved cautiously through the dark.

Sounds of the Red Ones came to them at times. And once they hid in a narrow branching cleft that came abruptly to a dead end, while a force of red warriors marched hurriedly through the passage they had just left. Back in their hiding place Rawson stood tense and ready, with his weapon till the last of the enemy was gone.

Always he was frantic at thought of the time that was slipping past—until, at last, the narrow passage that they followed cut transversely through another large runway that glowed faintly from some distant light.

With that first gleam of light there came over Dean Rawson an odd change. Something within him had been cold with fear. Fear of the flying minutes. Fear that Loah might have lost her way in this tangled labyrinth of winding ways. And now, suddenly, he was care-free, filled with an absurd joy. Nothing mattered. They were to die, but what of that? Loah had chosen death; he would see that when it came to her, it would be quickly and without pain. And as for himself, if before he died he could remove this ruler of an enemy race....

So when Loah leaned close and whispered, "The light—it shines from the council room of Phee-e-al," Dean replied almost gaily; "I've got to hand it to you—you sure do know all the back alleys." Then he stuck his head cautiously out into the dimly-lighted corridor.

It was broad. He saw where their own little passageway went on from the opposite side. But the light—the light! At his left, not a hundred steps away, was a room, brilliantly lighted. And across it, in gleaming splendor, stretched a low wall—a barrier of gold. It was the council room, where once before he had faced Phee-e-al in all that savage's hideous splendor.

He listened. All was silent. Then Loah whispered: "Phee-e-al comes this way when he goes to the council room. But when he comes, or how often, I do not know."

Dean pressed her back into the narrow way with his hands. "Wait here!" he said, and gave her the flame-thrower. "I've an idea!" He stepped softly out into the broad passage and on naked, noiseless feet, moved swiftly toward the lighted room.

It was empty. Beyond the barrier were no red figures, nor were there whistling voices to echo as he had heard them before. Here was the throne where Phee-e-al had sat; here the priests had stood; there, along the wall, were the chests.

Fully twenty of them, each eight feet long, they stood ranged along the three walls of that part of the room protected by the barrier. No two of them alike; all of them were oddly carved and studded with jewels.

The chests were ranged in a straight row a foot or more out from the wall. He crossed to them swiftly. About here was where that priest must have gone. He raised one of the heavy lids till the light struck within.

Bones! Only fragments of a skeleton, blackened by age; a necklace of teeth from some animal's jaw; worthless trifles for the mummery of the priests. Then, beneath them, he saw two great fangs, a foot in length. They were curved, sharply pointed and yellow as old ivory.

What was it Gor had said of legends that told of ancestors coming from the outer world? Rawson knew that he was looking at priceless relics of the tribe, at the tusks of man's long extinct enemy, the great sabre-toothed tiger.

But he had neither time nor thoughts to spare for marvels new or old — he must find his gun. Yet, even then, he wondered what undreamed-of treasures the other chests might hold — what jewels, what paraphernalia of ancient kings.

He must be silent! Perhaps the next great glittering container might hold the blue gleam of his gun. And this time as the gem-studded lid was swung upward and back to rest noiselessly against the rock wall, Dean could not repress the audible gasp that came to his lips.

His own pistol! He had expected to find the one weapon, but, instead, the chest was filled with all it would hold of rifles and side arms and cartridge belts, all mingled in one indiscriminate heap.

They were twisted, some of them, and bent; discolored, too, evidently by flames. On some the stocks had been burned off.

Rawson's hands were suddenly trembling. There was one rifle that seemed unharmed; he brought it out, and hardly heard the little clatter that it made among the other weapons. An ammunition belt — he slipped out a clip of cartridges, made sure they fitted his gun, and threw one up into the firing chamber. He was fumbling for more of the clips when there pierced through his tumultuous thoughts the realization that he was hearing sounds not made by his own suddenly clumsy hands.

Marching feet, whistling voices — they came from beyond the room's farther end, beyond the entrance through which he had once been brought a captive. He took one step back toward the broad tunnel, then knew there were others coming there.

There was no possible avenue of escape. He threw himself in one wild dive into the narrow space between the chests and the wall, and pulled himself forward under the shelter of the one back-turned lid. The rifle was still gripped in his hands.

By the sounds that came to him, he knew that the outer room had filled with red warriors, and that another smaller group had come scuffing from the passage where he had just entered. And, by the echoing cry of shrill voices that shouted, "Phee-e-al! Phee-e-al!" he knew that the ruler was near.

Then there were footsteps approaching the chest. A priest no doubt; shrill whistling told of his anger. The concealing cover was jerked outward and down, and Rawson, staring above him, saw not the coppery face that he had expected, but the hideous white visage of Phee-e-al himself.

For an instant the ruler of the mole-men stood half stooped in petrified astonishment, and in that moment Rawson dragged himself to his feet. No chance to use the gun — the other was upon him, his gripping talons tearing Rawson's bare flesh. In one flashing thought, Dean cursed himself for the uselessness of his weapon — he should have taken a pistol, an automatic. Then, body to body with the savage, he was dragged out over the chest.

He had been holding the rifle above him, as he struggled from his cramped quarters. The savage had grabbed him about the shoulders, but his hands were still free; they held the gun on high. And in the second when he found his feet under him, as Phee-e-al dragged him clear of the chest, Rawson brought the breech of the gun crashing down upon the pointed skull.

He felt the talons release their hold. The priests were rushing upon him. Phee-e-al, too, had been only momentarily stunned — he was springing. Then Rawson whipped the rifle down in line, and the clamoring shrieks that filled the room with tumult were drowned under another roar.

He saw Phee-e-al fall. Even then, through all the pandemonium within his own mind, he thrilled with satisfaction at sight of a little dot and a spreading stain above Phee-e-al's heart, where only bare skin had been before.

The next shot took the foremost of the priests. The others paused, hesitant for a moment, ranged out in an irregular line. Past them, beyond the golden barrier, Rawson caught a confused glimpse of a sea of red faces. Green flames were stabbing upward from their ready weapons. The priests were between him and them, and there came to Rawson in that instant, through all the chaos of fighting and half-formed plans, the knowledge that these priests were a living barrier that held off the flames.

He fired once more to check them, then sprang for the wide entrance of the tunnel. He fired again back of him, shooting wildly as he ran, then saw Loah as she came from her hiding place with the flame-thrower ready in her hand.

"Quick!" he gasped. "Get back!" Then, with her, he was running stumblingly through the dark.

There could be no escape; even while they fled he knew it. And yet they almost made it—though the end, when it came, was one that neither could possibly have foreseen.

They were following a wide passage, one of the countless thoroughfares of the Reds. It was deserted. Loah flashed her light freely. Ahead of them the passage turned. Just short of that bend was a rift in the rocks.

"There!" Loah gasped. "Turn there. It will take us back to the *jana*." But the words were followed by a flash of green from dead ahead.

The flames that made it came quickly after and a dozen of the red warriors were before them, the light of their weapons slanting just above Rawson's head. His rifle was half raised—they would at least fight to the last. Then he realized that the green death was not swinging downward.

From behind them, in the corridor through which they had raced, came a chorus of whistling shouts. Rawson whirled to find more of the red fighters, and again, though their hissing green flames were held ready, they did not descend.

A priest, copper-colored, shining resplendently in the weird glow, detached himself from the group and stepped forward under the protection of their weapons. Loah's hand was depressing the muzzle of Rawson's rifle. "Wait!" she said. "He wishes to speak."

The priest stopped and addressed them. Loah answered; and to Rawson it seemed horrible that her lips and throat should be called upon to form those whistling words. Then she turned toward him.

"He says they will not harm you now if you surrender. Later, when they select a new ruler, he may order you set free."

Rawson was doing some quick thinking. The priest was lying, clumsily, childishly, but it might be he could bargain with them.

"Tell them this," he ordered Loah: "they are to let you go free—let you go right now! If they do that, I'll lay down my gun. If they don't, that priest will die before they get me. I don't think you can make it," he added, "but go back to the *jana*. Don't stop for anything. Drive it as fast as you can; you may still get there before Gor does his stuff. And take the flame-thrower in case you are followed—" He stopped; Loah was laughing.

"Did you really think, Dean-San, that I would desert you?" Again she laughed softly—laughing squarely in the face of that waiting death, a laugh that was half a sob, that caught suddenly in her throat as she stared at Dean.

He could not read the look in her eyes as their expression changed. "Yes," she said slowly, "yes, you are right. If I stay we both die, quickly."

Again her voice made whistling sounds; the priest replied. Then Loah threw her arms around Dean and kissed him. He was gripping his rifle; before he could take her in his arms, she was gone. She walked swiftly, the flame-thrower in her hands, toward the dark cleft in the rocks, through which she disappeared. And Dean, though she had done what he really wished, felt that all of his life and strength had gone with him with that fleeing figure.

He placed his rifle on the floor and, straightening, held out his empty hands; the priest's talons were upon his flesh.

"But I got Phee-e-al, anyhow," he was thinking dully.

CHAPTER XXV
SMITHY

Scarcely more than a vault in the solid rock, the room where Rawson lay. He had seen it for an instant when the priest, after tying his hands behind him, had hurled him viciously into the room. It had but one entrance, though up high on one wall was a crack some two feet in width that admitted fresh air. A little room, only some twenty feet square; but he would not suffocate — the priests did not intend that he should die — not yet.

He saw one of the giant yellow workers bring a big metal plate. He put it before the doorway; then, by the red glow, he knew that they had sealed him in.

"I got Phee-e-al," he thought. "I did that much to help. That may put a crimp in their plans, check the invasion up above. But Gor didn't do as I told him, or it didn't work. The twenty-four hours must have gone by."

Then, even in that thought, he found happiness. "That means that Loah is safe," he told himself. "The shaft is clear; she's on her way back right now."

He pictured the *jana* falling swiftly through that dark shaft. He saw in his mind the beautiful figure of the girl, lithe and slender, standing at the controls.

About him was a silence like that of the grave; his blood pounded in his temples like a throbbing drum. It was some time before he knew that, with that throbbing, other faint sounds were mingled.

They came from the wall beside him, sharp tappings muffled by distance, the faintest whispering echo of rock striking upon rock. *Tap-tap ... tap.* A longer pause.... *Tap.* They were making dots and dashes that blurred with the beating in his own brain.

In that dreadful silence he strained every nerve in an agony of listening. There was nothing more.

He had been roughly handled by the savages. His whole body was bruised and aching, his thoughts hazy and blurred. "Woozy," he told

himself. "Guess the old bean must have got a bad crack. Hearing things —
mustn't do that."

Again he tried to picture the girl, speeding on toward that inner world.
Was she thinking of him? Surely she was. He could hear her calling his
name. "Dean," she was saying. "Dean-San." The words were repeated, an
agonized, ghostly whisper — repeated again, "Dean-San — oh, Dean-San,"
before he knew that the sound was coming from overhead. Then a light
flashed once in the little room, and he saw her face, looking down.

She was beside him an instant later. "Dean-San," she was saying, "did
you think that I really would leave you?" She was pressing her lips to his.
Uncovering her light, she worked frenziedly at the metal cords that bound
his wrists, pausing only to repeat her caresses — and at last he was free.

"I reached the *jana*," she told him in hurried whispers, "and then I came
up. Their great room, where the Pathway to the Light begins, was deserted.
With a cord I pulled the lever, and the *jana* vanished. I could not leave it for
them to use. Then I followed — I knew by the sounds where they were taking
you. And now, what can we do, Dean-San? Where can we go?"

It was real! Loah was there beside him; he had her in his arms, his
bruised, bleeding arms whose hurts he no longer felt. And then, through his
mind, flashed the question: if this was real, what of the other — the rappings
he had heard? Perhaps it hadn't been a dream.

He lifted a fragment of rock and crashed it against the wall from which
those rappings apparently had come. Laboriously he spelled out his name,
remembering the dots and dashes from earlier flying days when planes had
been equipped with key-senders. He spelled it slowly and waited, while
only the silence beat upon him and the blood pounded in his ears. Then he
heard it. The answer came from a quicker hand:

"Rawson — this is Smithy."

But Smithy was dead! What could it mean? Slowly Rawson pounded
out the letters of his question: "Where — are — you?" The answer dispelled
his last doubt as to the reality of what he had heard.

It *was* Smithy. Others were with him, for Smithy said "we," and they
were prisoners, sealed up in a living tomb. But where? Smithy did not
know. He knew only that they were in a big room where the rocks had been
shattered and molten gold spilled on the floor. There was a hole in the roof,
but too small to get through — a round hole, about eight inches in diameter.
And, at that, Rawson interrupted to tap out a single word.

"Coming!" he said, and turned toward Loah and the light.

The girl had found a metal rope in her wanderings; she had used it to let herself down into the cave. And now it was she who helped Dean to pull his bruised body up and into the narrow crack. Loah had clung to the flame-thrower; they found it where she had left it up above.

The tapping rocks she could not understand, but she knew Dean had a definite plan in mind when he whispered: "The room where you first found me — do you remember? Do you know the way?"

"I will always remember," she said simply. "And, yes, I know the way."

Rawson caught glimpses now and again of that broad thoroughfare along which he had once traveled, a prisoner of the mole-men. But Loah knew other and seldom-used passages that roughly paralleled it; and then, after a time, Rawson himself knew in what direction they must go.

He knew, too, that they had followed a circular route, and that the room in which he had been sealed was not a great way from the place in which Smithy was a prisoner. Yet this had been his only way to reach it.

When they came to a sudden sharp turn, he realized that they were close. Beyond that bend would be the branching, lateral tunnel that led to Smithy's prison.

The main runway had been deserted by the Reds. Stopping often to listen, starting at times into side passages at some fancied alarm, they had met with no opposition. But now, from beyond the angling passage, came the familiar shrillness of the mole-men's voices.

Again the two concealed themselves, but no one approached. "It's a guard we hear," Rawson whispered. "They're guarding that entrance where we must go. They're taking no chances on Smithy's escaping." Then he crept to the point where the passage turned, the flame-thrower ready in his hand.

He drew back. For the moment it seemed to him physically impossible to turn this weapon upon them. They were savages, true, but it seemed horrible to slash living bodies with a weapon like this. Then he thought of the devastation those same weapons had wrought among the people of his own world. His momentary hesitation vanished. With one spring he leaped into the open where, a hundred feet away, red bodies were massed, and the air above was quivering with the green jets of their weapons.

His own flame-thrower he had turned to a tiny point of light; now it roared forth in fury as he swung it forward. They had no time even to aim their weapons or to turn them on. They were stampeded by the astounding attack. And still Rawson sickened as he saw them fall.

There were some who, panic-stricken, dropped their cylinders and leaped for safety in a narrow branching way. Rawson knew he should have killed them, knew it in the instant that they vanished, but that momentary, uncontrollable revulsion within him had stayed his hand.

He rushed forward now, Loah still bravely at his side — past the fallen bodies, through the choking odor of burned flesh. Grabbing up one of the weapons that had been dropped, he thrust it into her hands and said: "Wait here. Stand them off if they come back." Then he was rushing up the side corridor toward a room where once, in a far-distant past, he himself had been confined.

The flame-thrower lighted the way. It showed him the metal plate and the smooth, glassy rock that had been melted around its edge. He pounded on the metal and shouted Smithy's name.

Voices answered from within — voices almost unintelligible for the wonder and unbelief and joy that made them a confusion of wordless shouts. Then he stepped back and turned the blast of his weapon upon the rock at the edge of the plate.

The metal sheet moved at last, its top swinging slowly outward. Its base was held by the gummy, hardening rock. Then it broke free and crashed to the floor, and the light of Dean's weapon showed through the black opening upon the blanched faces of men, where eyes were still wide in disbelief.

Though they were looking at one of their own kind, it must have taken then a moment to realize that the naked body, clad only in a golden loin cloth, and the hands that held one of the fearful, green-flamed weapons, were those of a human. Then one of them broke from the others, sprang heedlessly across the still-glowing plate, and threw his arms about the barbaric figure.

"Dean!" he choked. "Dean, it's really you! You're alive!"

And Rawson's voice, too, was husky as he said: "Smithy, I thought you were gone. The radio said they had got you, old man."

Then other khaki-clad bodies, a dozen of them, were crowding through the hot portal, and Rawson came suddenly to himself.

"Quick!" he shouted. "They'll be after us in a second. Follow me."

Loah was waiting. Her own flame-thrower spat a little jet of green; it was the only light. Rawson saw here she had gathered up the other weapons and had turned them off so that even their little light would not blind her as she kept watch down the dark passage.

"Do we want them?" Dean shouted to the others. And Smithy echoed the question:

"Do we want them, Colonel?"

Colonel Culver, his face almost unrecognizable under its smears of powder stains and blood, snapped a quick answer: "No. We outrange them with our rifles. They're only flame-throwers, not ray projectors. Beat it! Run like the devil!"

Rawson snatched Loah's weapon and threw it with the others. It would be hard going, ahead—she must not be uselessly burdened. But he kept his own. Then with his one free hand he swept her up till she was racing beside him as they led the way.

"I should have kept the fire weapon," the girl protested; "I, too, can fight."

Rawson, speaking between breaths, reassured her: "Too heavy. Their guns will protect us—"

Behind them, a man's voice cried out once, a single, hoarse scream of agony; then the rock wall took the sharp crackle of rifle fire and threw the sound into crashing, thundering echoes.

CHAPTER XXVI
POWER!

A girl whose creamy body was strangely unsoiled by smoke or grime, whose jeweled breast-plates flashed in the light of her torch while the loose wrappings about her waist whipped against her as she ran. And Rawson, naked but for the golden loin cloth, running beside her. Then Smithy, and ten others in the khaki uniform of the service—it was all that was left of the fifty who had dared the depths. And now all of them were harried and driven like helpless animals in the burrows and runways of that underworld.

But not entirely helpless. Colonel Culver had been right: their rifles outranged the flame-throwers. And Rawson, looking past that first burst of rifle fire, saw the one flame that had reached them whip upward as its owner fell. Others of the Reds came crowding in after, and the jets of their weapons made little areas of light as they crashed to the floor. Then Colonel Culver took charge of the retreat.

Ahead of them and behind them was impenetrable darkness; only the nearby walls were illumined by the torch that Loah had been forced to turn on. And out of that darkness at any moment might come devastating flames. Culver detailed two men as a rear guard and two others to run ahead a few paces in advance. At intervals of a minute or two their rifles would crack, and the echoes would be pierced by the whining scream of ricochets, as their bullets glanced from the walls.

"We may not need them up ahead," Culver shouted to Rawson. "I don't understand it. The place seems deserted—there were plenty of them here before!"

"They've got something else to think of," Rawson shouted in reply. "I killed Phee-e-al—he was their leader. But they're after us now. They'll be running through other passages, cutting in ahead of us."

The tunnel turned and bent upward. For a full half mile they ran straight in a stiff climb. Between gasping breaths Colonel Culver shouted hoarsely: "Won't it ever turn? If they bring up their damned heat-ray machines they'll get us on a straightaway like this!"

Then Smithy's voice outshouted his with a note of hope: "We're almost there; I remember this place. There's where we mounted the searchlight. They've ripped everything out. Up ahead, one turn to the right, then a quarter mile, then a turn toward the crater. That runs straight for a mile, but there's a field gun at the bottom of the volcano. We'll be safe when we're on that last stretch."

Ahead of them the rifles of the two who ran in advance crashed out in a fury of fire as a green glow appeared. But this time the flame did not die; and Rawson, staring with hot, wide-opened eyes, saw that the ribbon of green swept transversely across the tunnel.

He could hardly stand when he came to a stop. Beside him Loah was swaying with weariness. The walls echoed only the hoarse, panting breath of the men. Then they crept slowly forward, where the passage went steadily up. Loah's light was out; she had slipped the cap on the torch at the first sight of that green.

They stopped but ten feet short of the deadly blaze. From a narrow rift in the left wall it streamed outward, the rock at the edges of that crack turning to red at its touch. It beat upon the opposite wall, where already the stone was melting to throw over them a white glare and the glow of heat. And, like a shimmering, silken barrier, whose touch could mean only instant death, it reached across the wide tunnel at the height of a man's waist and moved slowly up and down. The heaviest armor plate ever rolled could have formed no more impenetrable a barrier.

"And we almost made it," said Smithy slowly. "Look, beyond there — another hundred feet. There's the bend in the tunnel, a sharp turn — and we almost got around!"

Rawson reached for Loah's light. In the wall where the flame was striking, only a dozen steps back, he had seen another dark mouth, a ragged crack in the rock. He sprang to the entrance; it might be there was another way around. His first glance told the story, for he saw the walls draw together again not a hundred feet off.

"A blind alley," he groaned.

One of the two who had been their advance guard snapped his rifle to his shoulder. He was aiming at the glowing crack where the green light was issuing.

"A ricochet," he growled. "It may go on in and mess 'em up." But there was no whine of a glancing bullet that followed his shot; the softened wall had cushioned the impact.

Another man sprang beside him. He was shouting at the top of his voice while one hand reached into a bag that hung at his waist. "Get back, everyone," he said. "If I miss...." He did not finish the sentence, but pulled the pin from a hand grenade, then took careful aim and threw.

It went high — thrown there purposely; he had not dared aim it into the flame. But it struck the crevice fairly, and they heard it rattle on inside. The next instant brought the crack and roar of its explosion.

Like a winking signal light the green barrier vanished. Where it had been was only blackness and the dying glow of molten rock. Then, a hundred feet beyond, up close to the roof, the bend of the tunnel turned red; it seemed bursting into flame. Far back of them, down the long sloping way where they had come, shrill voices were screaming — and still there was no green flame to account for that tunnel end flaming red.

Rawson stood motionless. Loah, and the others beside him, seemed likewise petrified, until the voice of Culver jarred them into action.

"The ray!" he shouted. "It's the heat ray, damn them! Quick, jump into that cave!"

They had all retreated through fear of the grenade; they were opposite the black place into which Rawson had looked. Loah was close beside Dean; he threw her with all his strength into the black mouth of the cave, then he was one of a crowding, stumbling mass of men who followed after, and their going was lighted by a terrible torch of flame.

One man had stood apart from the others, farther across the wide corridor. His khaki-clad body flashed suddenly to incandescence, then fell to the floor. And inside the cave, where the walls came abruptly together to cut off any further retreat, Colonel Culver spoke softly.

"One more gone," he said. "That was Oakley. Well, he never knew what it was that hit him — and it looks as if we'll all get the same."

Through it all, Rawson had clung to his flame-thrower; unconsciously his hand had held fast to the bent handle of the cylindrical weapon. Now he set it down slowly upon the floor, then straightened his aching body laboriously.

Loah's light was still gleaming. He saw her eyes searching for his, half in terror, half in wonderment. Strange men with strange thundering weapons — he knew she was wondering if they still dared hope, wondering if these warriors of Rawson's race might be able to work further magic.

Dean put one arm tenderly about her and drew her close and his other hand came to rest upon Smithy's shoulder.

"It's the end, dear," he told the girl softly. "It's the end of our journey. You've been so dear and so brave. Pretty tough to lose out when we'd almost fought clear." Then, to Smithy: "Loah came back to save me—refused to go when she could have got away and been safe."

Already the air was stifling. The tunnel beyond the mouth of the cave was hot, though only at its end, where the invisible ray struck the rock surface squarely, was there red, glowing heat. Rawson suddenly saw none of it. He was seeing in his mind the world up above, his own world of great, free, sunlit spaces. Suddenly he was hungry for some closer link, no matter how slight, to bind him to that world.

"What day is it?" he asked. "Have you kept track of time?"

Smithy looked at him wonderingly. "Yes," he said, then added: "Oh, I see. You want to know what day this is when we die. It's the twentieth, Dean"—he looked at the watch on his wrist—"just two o'clock, the afternoon of the twentieth."

Within him, Rawson felt a dull resentment. He was being denied even this last trifling solace. "You're wrong," he said sharply. "You slipped up on your count."

"It doesn't make any real difference," Smithy said. But Rawson went on:

"We left the inner world on the nineteenth. At noon on the twentieth Gor was to cut loose the flame-throwers, melt a hole in the floor of the ocean. But it didn't work. I had hoped I could wipe out the mole-men, turn a solid stream of water down a shaft for over six hundred miles. It would have gone through the Zone of Fire, come flooding up into the mole-men world and spread out all over down deep where it's hot. It would have hit the Lake of Fire—all that!"

"I don't know what you are talking about, Dean." Smithy's voice was intentionally soothing; he knew Rawson was talking wildly. "But I know I am right on the time. We've kept track of it every hour since—"

Rawson's talk had sounded like insanity in Smithy's ears. He would have gone on—he didn't want to see Dean Rawson go out like that—but now he stopped. The rock was quivering beneath his feet.

And now Rawson, with a wild wordless cry, threw himself toward the flame-thrower on the floor. His voice rose to what was almost a scream. "It's worked!" he shouted in a delirium of joy. "It's the end of the brutes!"

Then, in words which the others could not comprehend but which somehow fired them with his own emotion: "Gor has cut it loose! Water, millions of tons of it! The Zone of Fire—steam!..." He threw himself flat on the floor as close to the hot mouth of the cave as he dared go, and the green flame of his weapon ripped outward and up as he aimed it.

From the passage, where it sloped downward toward the source of the heat ray, the sound of shrill, whistling voices had swelled louder. The whole tunnel now glowed green from the flames of an advancing horde. They were bringing their ray projector with them, Rawson knew, not that its beam was visible, but the white, dazzling glow from the end wall where the tunnel turned was still there.

"Shoot above me!" Rawson shouted. "Don't stick your guns out into that ray, but aim as straight down the tunnel as you can. Keep 'em busy. Keep 'em from coming too close."

Above his head he heard the beginning of rifle fire as the men crowded close to aim at the opposite wall at as flat an angle as they could. The air grew shrill with the sound of ricochets as the bullets glanced, but still the enemy came on, as their screeching voices told.

His own weapon was aimed up above. The roof of the tunnel was rough and broken. He directed the flame against the top of a great black granite block. In one place it was fractured. If he could cut it off above, make it fall to the steeply slanting floor.... He worked the full force of the blast methodically along the line he had chosen.

The air of the tunnel had been blowing gently, but now it came in sharp gusts that whipped in through the mouth of the cave, while it brought an unending growl and roar like distant gunfire from deep within the earth. The breeze had swelled to a steady blast when the rock crashed down.

"But that's no use," Culver had shouted, when the deafening sound of its fall had ceased. "They'll melt it in a second with their ray." Even as he spoke the great mass of granite softened and rolled downward as the enemy shot their ray on its lower side. The heat of it struck blastingly into the entrance to their retreat, yet still Rawson kept on, sawing doggedly with the weapon of flame at other great blocks above.

Now that distant thunder grew hugely in volume, and again the rocks trembled beneath them. The wind in the tunnel grew suddenly to a wild blast. It brought to them from a thousand other passages, the shrill, demoniac shrieking of air that was torn and ripped on projecting ledges of rock. Mingled with it was the sound of voices that screamed in terror, and the echo of feet running in mad flight down the tunnel.

The mass of stone, that had been melting under the invisible ray, cooled to red, then to black. Outside, the tunnel, now a place of roaring winds, was lighted only by the single flame of Dean's weapon.

"They've gone!" Culver shouted. "The ray's off. Get outside! Now we'll run for it!" And, with the others, Rawson sprang to his feet and leaped out into the tunnel which was no longer a place of death.

He heard the sound of their hurrying feet and a voice that cried: "Look out for the turn—the rock's hot," but he did not look after them. He was standing squarely, bracing himself in the blast of air, still directing the flame upon a block that hung stubbornly and would not let go.

He knew that Loah alone stood near. He heard other feet; someone was returning. Then Smithy was upon him, almost jarring him from his careful pose. Smithy was shouting.

"Come back, Dean!" he cried. "Are you crazy? Don't you know they'll be after us again?"

Rawson sprang as the big rock let go. It, too, crashed deafeningly upon the floor and rolled sluggishly downward beside the high hummock of glass that the first rock had become. They bulked hugely in the passage. They were eight or ten feet high, reaching across from one wall to the other.

Above them was still a space of four feet; Rawson estimated it carefully while he looked at the ceiling above. Then he shook off Smithy's hand that was dragging at him and returned to the attack; for now, above the top of the barricade he had built, white ribbons of vapor were streaming. He had to shout to his utmost to make Smith hear above the shrill shriek of the blast.

"Steam!" he screamed into Smithy's ear. "Live steam! We could never make it—before we got to the top we'd be cooked to a pulp. I've got to block it, got to seal it off." A whole section of the ceiling tore loose as he spoke, and the wind raised its voice like the scream of a wounded animal—or the cry of an overwhelmed and stricken people—as it tore through the space that remained.

It whipped the molten drops as they fell and made of them a deadly rain. Rawson, staring through the clouds of hot steam that now wrapped him about, called to Smithy to take Loah to safety, and kept the flame where it should be—until at length the last aperture was closed, the last gap in the wall filled in. And even after that Rawson kept the flame still playing above that wall till he had melted rock and more rock that flowed down to make the barrier a single heavy, solid mass.

Steam was coming now from the narrow cleft where the green light had flashed out to bar their way. But that was simple, and he sealed the gap shut with his flame.

He was gasping. The radiant heat from that molten mass had been torture that his naked body could never have borne but for the desperate necessity that drove him.

Smithy and Loah were again beside him. "Now," he choked, "we can go, but if there are any cross passages I'll have to block them too."

"There aren't," said Smithy, and added: "I thought you were crazy. You've saved us all, Dean; we never could have made it to the top. That steam was getting hot—hot as if it had come right out of hell."

"It did," said Rawson. Then the flame-thrower fell from his nerveless hand. He was swaying; his knees were trembling with weakness when Smithy and Loah, on either side, took his burned arms tenderly and helped him on where the others had gone.

Colonel Culver and a rescue party met them halfway. The Colonel had seen his men safely to the bottom of the volcanic pit. Others had run from their station beside a field gun to meet them; then Culver had called for volunteers and had gone back. And now there were plenty of willing arms to help.

The big lift, with its platforms of metal plates, awaited them at the tunnel's end. There was room on it now for all who were left; there was no crowding of men's bodies as there had been on the downward passage. Rawson was stretched on the floor-plates, whose touch was cool to his tortured body. Loah was seated that his head might rest in her lap on that absurd little fragment of skirt. She bent above him, whispering brokenly: "Dean-San—my dear—my own Dean-San! We live, Dean-San. I can scarcely believe it, but I know that we live, for I still have you."

But Dean was able to stand when that journey was done. First, though, there were men who placed him carefully on a stretcher and carried him, when he commanded, to the crater's outer rim. On the ashy floor of the crater a big transport was waiting with idling motors, but Dean would not let them put him inside. He wanted to look out across the world, to see it in reality as he had seen it in his own mind when all hope was gone. He wanted to look out once more across Tonah Basin and let his eyes rest upon country he had known.

Loah and Smithy walked beside him, as the first-aid men carried him toward that distant rim. The rocks there were cleft—it was the place where he first had seen the inside of the crater's cup. There he had them put him down; and, with the help of Loah and Smithy, he got slowly to his feet. While they lifted him, he wondered at the sound in this desert world where no sound should be. A terrific rushing, an endless roar—and then his eyes found the clouds of steam.

Below him was the Basin, the tangled wreckage of his camp. And there, where the derrick had stood, was a tall plume of white. It did not begin close to the ground—superheated steam, until it cools and condenses to water vapor, is invisible—but a hundred feet above the sand. And, from there on up, two thousand feet sheer into the air, was a straight shaft of vapor, rolling up for another thousand feet into billowing clouds that the afternoon sun turned to glorious white.

"Power!" gasped Rawson. "Power—and it will be like that indefinitely!" Then he laughed weakly. "I had to go down there to do it, to make Erickson richer, but it was worth it. In there the ocean will slowly subside. Gor and his people will find their lost lands; the column of water in the shaft will hold the back-pressure of steam. And here, I have Loah, and that's all—but that's enough!"

He put one arm, still with the bandages of the first-aid men, about the girl. "I hope you'll be happy, dear," he said softly, and turned back. But Smithy barred the way.

"That isn't all," said Smithy jubilantly. "You see, Dean, Erickson fired you—Erickson thought you had run out on him. Instead of backing you up, he quit. So I bought them all out. Whatever is there, Dean—and it's worth more millions than I dare to think about—you own half of! Now get back on that stretcher. Just because you've saved all our necks up here on top of the earth, you mustn't think you can keep an Army ship waiting all day!"

www.ingramcontent.com/pod-product-compliance
Lightning Source LLC
LaVergne TN
LVHW042153190726
843493LV00006B/1655